Sustainability

MANAGEMENT

Sustainability MANAGEMENT

LESSONS FROM AND
FOR NEW YORK CITY,
AMERICA, AND THE PLANET

Steven Cohen

COLUMBIA UNIVERSITY PRESS
New York

Columbia University Press
Publishers Since 1893
New York Chichester, West Sussex

Library of Congress Cataloging-in-Publication Data

Cohen, Steven A.
Sustainability management : lessons from and for New York City, America, and the planet / Steven A. Cohen.
p. cm.
Includes bibliographical references and index.
ISBN 978-0-231-15258-7 (cloth : alk. paper) —
ISBN 978-0-231-52637-1 (e-book)
1. Sustainable development. I. Title.

HC79.E5C596 2011
338.9′27—dc22

2010048228

Columbia University Press books are printed on permanent and durable acid-free paper.

This book is printed on paper with recycled content.

Printed in the United States of America

c 10 9 8 7 6 5 4 3 2

To Donna

Contents

Preface

The 1975–1976 academic year was my first as a graduate student, and I took two courses that year that have influenced my thinking—and, in fact, my career—ever since. The first was a course in environmental politics taught by Lester Milbrath, with Sheldon Kamieniecki as his teaching assistant. The second was organization theory—in part, a course on public management—taught by Marc Tipermas. For the past third of a century, I have focused my research, teaching, and professional practice in the areas of environmental policy and management. My goal has been to use everything we know about organizations to make the implementation of environmental protection policy more efficient and effective. The idea of connecting environmental protection to organizational management seemed obvious to me.

Environmental policy in the twentieth century was dominated by a discussion of the trade-off between environmental protection and economic growth. I always felt that it was a false trade-off, but it was not until I learned about industrial ecology and the issue of global sustainability that I had a framework for understanding why that was so. What I learned was that wealth itself depended on a well-functioning biosphere. The effective delivery of food, water, and air—all biological necessities—requires a well-functioning set of ecosystems. The environment is not something pleasant to enjoy; it is something essential to survival. The great Amory Lovins said it well, writing about climate policy:

> A basic misunderstanding skews the entire climate debate. Experts on both sides claim that protecting Earth's climate will force a trade-off between the environment and the economy. According to these experts, burning less fossil fuel to slow or prevent global warming will increase the cost of meeting society's needs for energy services. . . . Environmentalists say the cost would be modestly higher but worth it; skeptics, including top U.S. government officials, warn that the extra expense would be prohibitive. Yet both sides are wrong. If properly done, climate protection would actually reduce costs, not raise them. Using energy more efficiently offers an economic bonanza—not because of the benefits of stopping global warming but because saving fossil fuel is a lot cheaper than buying it.
>
> (Lovins 2005, 74)

The question then emerges—with nearly seven billion people on the planet, how do we extract our needs from the planet without destroying it? The resources of the Earth seem quite fixed and finite. Do we know of any resources that do not expire when we consume them? Yes: the sun. The sun is a source of energy that is virtually infinite. Solar power stimulates photosynthesis and the regrowth of plant life year after year and seems to hold the potential to be a nonfinite energy resource. There might also be a form of nuclear power that could provide the possibility of infinite energy. The current form of nuclear power, while technically sound, creates siting and waste issues that are difficult to resolve.

In addition to moving away from finite resources such as fossil fuels and similar substances, industrial ecology preaches a religion of zero waste in production. A production process that emits useless effluents or emissions is inherently less efficient than one that does not. Just as Total Quality Management equated wasted labor or materials with poor management, pollution seems to be evidence of a wasteful and poorly managed production process (Cohen and Schonhardt 2008).

Could we grow our world economy, open up opportunity to the poorest of the poor, and manage to keep the planet intact? Could we manage to maintain the vibrant, dynamic lifestyles we invented in the developed world while keeping the planet productive and viable? These are the fundamental issues of sustainability management. In order to pose these as questions of management, this book divides sustainability into a number of operational, component parts. The idea of this book is to survey the concept of sustainability as it is thought of in a number of arenas, including energy, water, and food supply.

The book begins by defining sustainability management and asks if a sustainable economy is feasible. In this definitional chapter, I delineate the structure that will be followed throughout the entire book where I define a set of technical, financial, managerial, and political challenges. In the first chapter, I discuss the overall challenge of sustainability management. Chapter 2 addresses sustainable manufacturing and service business. Green business is in style, but once we get past the "greenwashing," how do we ensure that businesses actually operate sustainably? Chapter 3 discusses energy, the single most important sustainability issue we face. Renewable and carbon-free energy could have a transformative impact on the sustainability of our economy. Water filtration, desalinization, food production, and a range of other material transformations become far less expensive if energy costs are low. Environmental impacts can be reduced and global warming could be eliminated as well.

Moving directly into the issues of biological necessity, the fourth chapter assesses water sustainability. Water is largely an issue of distribution and preventing degradation and contamination. It intersects with the energy issue, because of the cost of processing water, and with the climate issue, because of changes in patterns of rainfall. A similar and related issue is that of food supply, addressed in chapter 5. Throughout human history, most people were directly involved in growing or gathering their own food. Today, for the first time, most of the world's population is urban, and food production has become an industry. The use of technology to expand food production is a given. The world's population depends on the technology of food production and distribution for survival. The questions we will ask in chapter 5 are: which technologies are least damaging to the biosphere, and how do we maintain food production that can be sustained over the long run?

In chapter 6, we move up a level in the focus of our analysis to the interdependent systems in which we live: our cities. In 2007, for the first time in world history, the majority of people on the planet lived in cities (Moreno and Warah 2006). Cities present opportunities for efficiencies because they provide economies of scale and economies that result from population density. However, their survival requires the operation of a large number of complex, interdependent systems. Dwellings and other structures must be sited and constructed. Energy, water, and waste and sewage removal must be provided to each building. Air quality must be maintained, and food supplies cannot be interrupted. Commerce and transportation must be facilitated, and all of this must be provided at a cost that allows the city to attract businesses and residents. In this chapter and throughout the book, I take many of my examples from New York City. I do this for two reasons: first, because it is one of the most sustainable and sustainability-minded metropolises, and second, because it is the city that I experience daily and know most about. I believe that many of the lessons learned in New York are applicable in other places.

In chapter 7, I discuss the issues that must be addressed if the entire planet is to survive. These include the preservation of biodiversity and ecosystems as well as the reduction of global warming and ensuring that our air, land, and water remain free of poisons. In addition to these environmental issues, we now have the technical ability to destroy many of the planet's life forms, and so issues of war, terrorism, and human conflict are also central to sustainability. The final chapter of the book takes all of the issues we have addressed one at a time and tries to ask the overall question of global sustainability: can we manage this complexity, or will the modern economy destroy this planet? Of all of the issues we have reviewed, which are most important? What are the key technical, financial, organizational, and political challenges we face? Can we build a sustainable economy? Can we manage our organizations according to the precepts of sustainability?

The massive BP oil leak in the Gulf of Mexico that began in April 2010 brought the problem of sustainability management to the top of the political agenda in the United States. Our continued dependence on fossil fuel, along with lax regulation, inadequate technology, and incompetent management, led to an ecological catastrophe of epic proportions. The need for more sophisticated sustainability management was made obvious by this disaster, and the purpose of this book is to provide a resource that might facilitate the training of such managers. I strongly believe that within a generation, all competent managers will be sustainability managers: people who know finance, performance management, human resources, marketing, strategy—all of the techniques of modern management—and also have a deep appreciation of the physical dimensions of sustainability. A well-rounded, competent manager will know about energy efficiency, ecology, green architecture, and hydrology, as well as a number of other environmental fields.

Throughout the book, I draw frequently on my own experiences and case material from New York City. It is not that I think that New York City is a typical American or world city; rather, I believe that my home town has many positive and negative lessons for students of sustainability management. I know that—like most New Yorkers—I am hopelessly parochial. Like the famous New Yorker magazine cover, I really do think that (or at least act as if) the Hudson is as wide as the Pacific. While I suspect that I may have gotten that wrong, I genuinely do hope that you find these examples helpful in detailing the core concepts of sustainability management.

Acknowledgments

A book like this represents the work of many people, and I want to acknowledge some of them here. Laura "Lilly" Kelly, a recent graduate of Columbia's MPA Program in Environmental Science and Policy, did a terrific job helping with many of the sustainability memos that appear in the volume and assisted in research throughout the volume. Andrea Schecter of the Earth Institute did a wonderful job in helping conduct some of the research for this book. Eve Solomon, a recent graduate of Barnard College and a researcher at the Earth Institute, played an invaluable role in research and editing. Sara Schonhardt, a graduate of the Masters of International Affairs Program at Columbia, worked on several of the pieces that originally appeared in the *New York Observer*. Being at Columbia gives me the opportunity to work with spectacular students, and I am grateful for their help on this book.

In addition to the students I get to work with, I also want to thank a number of Columbia staffpeople who contributed to the development of this volume. Youngmi Jin, my former executive assistant and now an Earth Institute financial analyst, helped me at the start of this project, and then my special assistant, Allison Laude, helped me as we worked to complete it. Paul McNeil, vice dean of Columbia's School of Continuing Education, convinced me to develop and direct a new masters program in sustainability management. Louise Rosen, a key senior manager at the Earth Institute and associate director of the new masters program, was his partner in creating this new program. I thank her for her careful critique of the manuscript. Once I agreed to direct and teach in the sustainability management program, this book became an immediate necessity. I now teach a course in sustainability management, and this book will be the main text in that course.

Throughout the book, readers will find case studies on some major issues in sustainability management that are based on columns I wrote for the *New York Observer* online edition as well as blog posts for the *Huffington Post*. I appreciate the opportunity to modify and use these items in this book.

The intellectual debts I owe in developing this book are too numerous to mention, but I must at least acknowledge the most obvious. It starts with my boss, Jeff Sachs, the director of the Earth Institute and the person who taught me about sustainable development. I must also acknowledge the entire Earth Institute faculty: Peter Schlosser, the founding chair of that faculty, and our colleagues: Scott Barrett, Wallace S. Broecker, Mark A. Cane, Robert Chen, Joel E. Cohen, Peter Coleman, Patricia Culligan, Ruth Defries, Richard J. Deckelbaum, Peter B. deMenocal, Michael Gerrard, Steven L. Goldstein, Joseph Graziano, Geoffrey M. Heal, Patrick L. Kinney, David H. Krantz, Klaus S. Lackner, Upmanu Lall, Robert Lieberman, Edward Lloyd, Vijay Modi, John C. Mutter, Shahid Naeem, Robert Pollack, Stephanie L. Pfirman, Richard Plunz, Kenneth Prewitt, G. Michael Purdy, Pedro Sanchez, Elliott Sclar, Elke Weber, Stephen E. Zebiak, and Allen Zweben. I also owe an intellectual debt to Michael Crow, the founder of the Earth Institute and the force behind the School of Sustainability at Arizona State University, and his right hand during those formative years, Lewis Gilbert, now at the Nelson Institute of the University of Wisconsin.

In any work I do in any form of management, I must always acknowledge my debt to Marc Tipermas, who introduced me to the field of organization theory in graduate school and then hired me at the Environmental Protection Agency at the start of my career. In recent years, I have again had the pleasure of working with Marc at Willdan, a terrific consulting firm with a large practice in sustainability and energy efficiency. I must also thank my long-time partner, friend, and coauthor, Columbia University's Bill Eimicke, the best management teacher, consultant, and translator I have ever known. I also thank Ron Brand, former director of the EPA's Office of Underground Storage Tanks, who introduced me to the work of W. Edwards Deming in quality management and taught me that the goals of environmental protection and management innovation could be mutually reinforcing. In the field of environmental policy, I must thank my long-time friend, colleague, and coauthor Sheldon Kamieniecki, dean of social sciences at UC Santa Cruz, and my newer colleague Tanya Heikkila, now at the University of Colorado at Denver. All of these people shaped the ideas that fill this brief volume, but only I am responsible for the mistakes and omissions you will undoubtedly discover.

Last, but not least, I continue to be grateful for the love and support of my family: my wife, Donna Fishman, to whom I dedicate this book; my children, Gabriella Rose and Ariel Mariah; my parents Marvin and Shirley; my brother, Robby; and my sisters, Judith and Myra.

Sustainability

MANAGEMENT

Chapter 1

What Is Sustainability Management?

Sustainability Management Defined

No book about sustainability should begin without reference to the definition of sustainable development that originated at the 1987 Commission on Environment and Development, also known as the Brundtland Commission. That commission defined sustainable development as "development that meets the needs of the present without compromising the ability of future generations to meet their own needs" (World Commission on Environment and Development 1987). MIT's Richard Locke, one of the founders of that university's terrific Laboratory for Sustainable Business, uses the image of a piece of fabric to define sustainability:

> I build on the Brundtland Commission's definition of sustainability, which focuses on using resources today in a way that ensures there'll be resources to meet the needs of future generations. . . . Climate, environment, energy, social standards—they're all linked. One of the metaphors we've used a lot over the last couple of years is to think of sustainability as a fabric. You pull a thread and everything comes together.
>
> (Locke 2009)

Sustainability management is simply the organizational management practices that result in sustainable development. In the modern industrial world, sustainability management is the practice of economic production and consumption that minimizes environmental impact and maximizes resource conservation and reuse.

That is the basic definition. What does it really mean? How can one draw the line between management for sustainability and management that does not lead to sustainability? At the heart of sustainability management is a concern for the future. Most common management practices focus on the present or, at best, a one-year planning cycle that relates to an organization's budget and fiscal year. As J. Ehrenfeld (2005) has observed: "The problem really stems from management's failure to see unsustainability as a deep-seated systems failure and to appreciate the extent to which radical thinking and action are required to embark upon a sustainable trajectory." As performance measurement systems have become ubiquitous within organizations, management has focused on reporting cycles that include quarterly, monthly, weekly, and even daily reports. This focus on the present creates an organizational culture and environment that makes it very difficult for the issue of long-term sustainability to be taken seriously.

Issues of sustainability can be difficult to define and operationalize. According to Ehrenfeld (2005), "managers must evaluate critically the core values and mission of their business in terms of both the unsustainability *and* the sustainability they create. Reducing unsustainability is not the same as creating sustainability."

Why wed one complex concept, sustainability, to another complicated concept, organizational management? In part, I do this because my goal is to add the notion of sustainability to the definition of effective management. Organizations seek to maintain themselves. An organization that fails to take into account the long-term sustainability of the planet may survive while everything around them dies, but the odds are against them. I will argue (since I can't prove it) that healthy organizations depend more than they think on a healthy planet.

This comes down to the issue of waste and the relationship of efficiency to good management. Why wouldn't an organization strive to maximize the productive benefit of all of the resources that it has access to? As the sustainability scholar Richard Locke has noted:

> Efficiency (lower unit costs), quality, reliability—often these "positive" attributes of companies go hand in hand, managers will tell you. Now think about sustainability. If a company is good at de-

> veloping systems that deal with health and safety, and/or treating waste and water, and/or devising innovative ways to reduce energy consumption, and so forth, they usually have their act together on many other, fundamental, how-they-do-business fronts. In other words, companies that have thought hard about how to establish various management systems that promote more sustainable business practices are also companies that have thought hard about how to be more efficient or innovative or differentiate their products and services in the market.
>
> (Locke 2009)

One way that successful organizations thrive is by keeping the costs of production and service delivery as low as possible without sacrificing quality. If there is a technology that can allow you to use less energy, water, or other materials in production, all things being equal, why wouldn't you use it? The issue is often one of competing capital investments. The funds for reducing waste are the same funds needed to actually produce the product or service you are selling. Shouldn't the rate of return for sustainability investments be analyzed the same way you would analyze other investments? Professor John Sterman of MIT expressed this when he observed:

> Another big impediment is that there's a fundamental worse-before-better tradeoff. If you want to redesign your operation to use less energy, use fewer inputs, produce less waste, it's likely to have a positive return on investment, but like any investment, in the short run performance will suffer. This goes beyond the classical "you have to invest so your cash flow is negative first and then becomes positive later," although that's part of it. There's a much deeper issue there, which is reorganizing, redesigning processes, investing in process improvement. Doing all that work is disruptive in the short run.
>
> (Sterman 2009)

The mania for short-term financial gains can inhibit the implementation of new sustainable energy sources, many of which require large financial investments but do not produce returns until years later. Cheap credit encourages long-term investment, while a high interest rate skews the cost-benefit analysis, makes future returns look less promising, and encourages short-term investment. One study shows that the levelized costs for photovoltaic solar-cell energy reach $215 at a 5 percent discount rate and soar to $333 when a 10 percent discount rate is applied (Renewable Energy Focus 2010). The rate of discount is one of the leading factors in determining which renewable energies will be adopted. The government decides upon the discount rate, which was 3 percent during the Bush administration and is currently .75 percent under Obama. By setting a low discount rate, government endorses investment in the long-term, shifting the focus away from the short-term.

Administrations can provide financial incentives for investment in renewable energy by lowering their discount rates. A study conducted by the Organization for Economic Cooperation and Development finds that low discount rates are associated with investment in low-carbon technologies such as nuclear energy, while high discount rates are associated with investment in coal without carbon capture and gas-fired combined-cycle turbines (OECD 2010). The study concludes, "it is evident that interest rates and hence the discount rates investors use . . . have a major impact on the absolute and relative costs of investments in power generation" (OECD 2010, 158). In order to make the switch to sustainable energy sources, we have to think in the long run. Therefore, "the future should not be discounted too steeply. Ensuring a stable investment environment with low real interest rates is indeed one of the most effective steps to ensure sustainable development in the electricity sector and beyond" (OECD 2010, 161). Forcing ourselves to keep discount rates low will allow future generations to enjoy the benefit of sustainable energy sources.

In 2008 and 2009, we learned that thinking in the short term can be the enemy of a sound economy. This was clearly articulated by Mindy S. Lubber, the President of Ceres, a U.S. coalition of investors and environmental leaders, in mid-September 2008:

> The fiscal crisis on Wall Street is a painful lesson in how entire industries can delude themselves into ignoring the most fundamental issues—in this case, the hidden risks from easy sub-prime mortgages. It also reveals the vast pitfalls of an economic system obsessed with short-term gains and growth at all costs while ignoring essentials such as building long-term shareholder value and

> protecting the future of the planet. As we confront global climate change—perhaps the biggest challenge mankind has ever faced—business and government leaders have an opportunity to learn from the ongoing Wall Street debacle and get it right.
>
> (Lubber 2009)

It could be that issues such as the safety and well-being of workers and the sustainability of the planet should be subject to a different sort of analysis. While government and nonprofit organizations are designed to facilitate alternative metrics for allocating capital resources, private organizations are not well suited to allow lower rates of return on capital to achieve organizational goals.

Is It Feasible? Can the Impact of These Practices Be Great Enough to Permit True Sustainability?

With the population of the planet growing toward and past seven billion and probably to a peak of ten billion (U.S. Census Bureau 2008), it is reasonable to ask whether sustainability is feasible. Is there enough capacity to produce the food, energy, water, air, and other biological necessities that we require for human life? In addition to the absolute requirements to keep our organism alive, there is also the issue of quality of life. One could imagine a planet that allowed life but with such a high level of disease and discomfort that its quality would be substantially reduced.

While it will not be smooth or simple to build, I believe we are at the start of a sustainable or green economy. My reasoning here is not simply naive optimism but the recognition of necessity. The false wealth of the period ending in 2008 and 2009 focused many of us on the need for a solid, understandable basis for our economy. One part of a solid economy is found in free-market capitalism, where investors risk their wealth to create a valued product or service. The success of this enterprise produces wealth, and some people get rich while others do not. Along with capitalism comes the recognition that a certain amount of income inequality is not only acceptable but also desirable.

The question is: how much inequality should there be? The answer is that the level of inequality cannot be so great that people on the bottom of the ladder cannot live a decent life. Inequality must not be so high that there is hunger, hopelessness, untreated disease, violence, and inadequate access to education. We've learned that a large middle class makes societies wealthier and can contribute to political stability. But without public policy to encourage a middle class, the logic of the unregulated market leads to greater and greater inequality. This, in turn, leads to economic and political instability that can threaten the peace and security of the social order. The second part of a solid economy is the creation and maintenance of production and wealth over the long term. A concern for the long term is central to the definition of sustainability.

If a nation achieves wealth by oppressing its people or damaging ecological resources, it eventually pays a price for its misdeeds. In the United States, we paid the price of oppression under slavery with a brutal civil war and its racist aftermath. We have also spent hundreds of billions of dollars to manage and clean the poisons we have released into the environment and still release in the name of industrial production. China has only started to learn the environmental and financial cost of rapid development. In the end, they will pay, and here in the United States, we will continue to pay as well. Short-term gains are often bought at the price of long-term pain. This is a concept that is gaining currency. Landing on a carrier in a pilot's outfit does not mean you accomplished your mission. Sometimes a fund that pays off the same high return year after year turns out to be an unsustainable Ponzi scheme. On the other hand, an experienced pilot who knows his stuff and is humble and dedicated just might manage to land a jet plane on a river. Most people can distinguish solid from shaky. Sustainable means solid, dependable stuff that is designed to last for the duration.

What do we need to develop a sustainable planet? There are a number of prerequisites:

Reduce the destructive elements of competition among people and nations.

End the growth of the human population, end poverty, and eliminate extreme levels of income inequality.

Develop a renewable economy not based on fossil fuels.

Learn how to reduce the damage we do to our environment.

Peace

With the presence of weapons of mass destruction, we need to develop a system of international law that reduces the probability that these weapons will be used. Our current system of international law, balance of power, and diplomacy has failed from time to time, but it has at least prevented unimaginable disaster from taking place. We need to improve these international institutions. Unfortunately, as destructive technology becomes more lethal and the world's population more urban, the probability of catastrophe increases. The technology of law enforcement is also improving, but the constant threat of terrorism and the steps needed to combat these threats can reduce freedom and impair quality of life.

Population and Poverty

The human population continues to grow. In 2009, the world's population grew by about six million. This growth was uneven across the globe. In developed countries that do not encourage immigration, such as Japan, the population has been declining. In 2008, Japan's population decreased by 150,000 (U.S. Census Bureau 2010). In the developed world, population growth would end if not for immigration. In developing nations, the population is still growing. The reason for these different growth patterns is simple. In the developing world, parents cannot be even partially confident that their child will grow to be an adult, and in the absence of social security, children are the best form of old-age insurance. Moreover, in an agrarian world, children are needed to grow and harvest food. In the developed world, children are typically economic liabilities; they cost a great deal to raise and educate. We love and value our families, but we do not raise children for the economic benefits they bring.

People who study economic development and population talk about something they call a "demographic transition." This is what happens when a developing country makes the transition to full economic development. Children are no longer perceived to be economic assets but rather economic liabilities, and the population stops growing. The best way to end population growth is to end poverty (Sachs 2005).

Ending poverty also leads to sustainability in two other ways. First, poverty breeds political conflict. Without an ownership stake in society, people have less to lose and may be drawn to conflict. Parents who can provide for their children and realistically hope for a better life for them will favor peace over war (UNDP 2005). Second, some of the best brains that will one day invent a new technology or the cure for cancer may very well be trapped in a life of poverty and will never get the education they need to help us think our way to a sustainable future.

Energy

To reduce damage to the biosphere, global warming, and the cost of energy, we need to transition our economy to renewable, nonfossil fuels. While there are plenty of fossil fuels left on the planet, extracting those fuels will only become more difficult and expensive. Burning fossil fuels will continue to damage our ecology and atmosphere. Renewable energy is the key to the green economy. Without it, such an economy will never be achieved. The Obama administration's early energy initiative was a critical first step in developing this new energy economy.

The green economy aside, the development of renewable energy is proving necessary to human health, livelihood, and ecology. The 2010 BP oil spill in the Gulf of Mexico is evidence of the danger that dependency on fossil fuels presents and is a tragic indication that in the long run we will not be able to rely on oil for energy. The spill's economic, political, and ecological costs should convince corporations, public officials, and the public that fossil fuels will need to be replaced by renewable sources.

A few months after the spill, one reporter wrote, "as the catastrophic oil spill in the Gulf of Mexico continues to wreak havoc, renewable energy may never have looked better" (Choi 2010). A poll taken by Stanford University in June 2010 finds that while about three-fourths of the one thousand Americans polled oppose new taxes on gas and electricity, 84 percent polled favor tax breaks on renewable energy sources (Choi 2010). A *New York Times*/CBS survey reveals that Americans think the nation needs a fundamental overhaul of its energy policy, and most expect alternative forms to replace oil as a major source within twenty-five years, but they do not trust the government to make the necessary changes happen (Broder and Connelly 2010). It is clear that while we understand the high societal costs of a fossil fuel economy, Americans are unsure about how to make the switch to a green energy economy.

Ecological Footprint

The year 2007 was a turning point in world history—for the first time, a majority of the world's population lived in cities (Moreno and Warah 2006). One of the great paradoxes of modern life is that given the size of the world's population, it is better for the planet's ecosystems if people live together in cities than if they are dispersed throughout the countryside. By living in cities, we make it easier to preserve natural environments outside of cities. Densely populated New York City is much more energy efficient than most other places in the United States. Judith Layzer calls Manhattan an "ecotopia." As we learn to manage our energy, water, and waste more effectively through increasingly sophisticated technology, we can reduce our impact on the planet and gradually transition to sustainability (Layzer 2008). Layzer argues that cities have an important role to play in achieving environmental sustainability but that they can't do it without the help of national governments.

Can we do it? Can we get from here to there? Let's put it this way: if we don't learn to grow our economy while protecting our environment, we may survive, but, to paraphrase Nikita Khrushchev, the living will envy the dead. While the human species has some irrational tendencies, we don't tend to be suicidal. The opposite of sustainable development is short-term wealth that can't be maintained. This all sounds a little like Wall Street at the start of the twenty-first century. Still, I like to think we are a teachable species and that sustainability is actually feasible. While I am optimistic, it is an open question. The 2010 BP oil spill was a sustainability management failure of epic proportions. That event calls into question our ability to manage the complexity of economic life on this fragile planet.

The Technical Challenges of Sustainability

If we are to achieve worldwide economic development while maintaining a functioning biosphere, we must learn to control the impact of our activities on the planet. We need to learn to distribute, process, and efficiently use water. According to Layzer (2009): "The single biggest impediment is the fact that none of the things that are limited in our natural system have prices. We don't price carbon, we don't price ecosystem services. If we're really going to do this—I mean if we're *really* going to do this—then we need to put a price on what's scarce." While I do not agree that pricing is the only way to make public policy, the market is a powerful influence on human behavior and an important tool—even if it isn't "the single biggest" one.

Sewage and other waste must be cleaned before it is returned to our rivers, lakes, and oceans. Food must be mass produced while retaining the capacity for regrowth and regeneration. Our energy supply must be based on the virtually limitless source of the sun or a more politically and environmentally acceptable form of nuclear power. We must become more conscious of and careful with the complex web of life that supports the existence of species other than our own.

To accomplish all of these tasks, we must dramatically improve our understanding of this planet and the impact of our actions on the biosphere. The first step in developing the technology of sustainability is to develop the means of measuring the health of the planet in all of its intricate dimensions. The goal of sustainability is nothing less than planetary management, which is an audacious goal that we are a long way from achieving. Measurement is a critical element of management. If you can't measure something, you can't manage it, because without measurement, you cannot tell if your management actions are making conditions better or worse. The specific measures of the planet's conditions will help us identify problems and begin to work toward solutions.

Climate change is among the first planetary health problems that we have been able to identify. Measures of temperature, CO_2 concentrations, the thickness and extent of the polar ice belt, and rising sea levels have all been used to define the dimensions of this problem. We know there is too much carbon dioxide in the atmosphere, we know where it comes from, and we know that it is warming the planet. We don't know what impact the warming will have or how to adapt to it. The most fundamental technical challenge is sustainable, fossil fuel–free energy. Today, renewable energy is more expensive than fossil fuels. We need new technology to change that equation if we are to shift away from fossil fuels. Will that happen? The cost of renewable energy—the prices of solar power, wind power, and battery storage—will come down as the technology develops. Think of computers. The computer I am writing this on sits on my lap and is more powerful than the million-dollar-plus mainframes of the 1960s and 1970s. As mass markets are developed and technology is refined, prices come down, and today's infeasible ideas become tomorrow's everyday experiences.

How do we get this done? How do we get from here to there? In the case of computers, a lot of the basic research and development came from the Defense Department and NASA. Our rockets, missiles, and space capsules needed smaller, more powerful computers. And then there is the Internet, which was also developed by the government. Our military computers needed to communicate with one another, so the government allocated funds to develop a system that would link computers. One thing led to another, and eventually we had the Internet. Government paid the costs of development, and then the technology was turned over to the private sector—and a new industry was created.

Sometimes national security drives the development of technology, and sometimes public health is the motivating force. London, for example, developed sewers and indoor plumbing to prevent disease. New York developed a hugely expensive water-supply system because local sources were polluted. I'm sure someone was saying, "Do you know how expensive this indoor plumbing will be? We'll all go broke installing these pipes and pumps everywhere!"

More recently, we had some of the same arguments raised against paying the costs of installing air pollution devices on cars and power plants and against spending billions of dollars on sewage treatment plants. We did all of that, and the economy continued to grow. In fact, the economic benefits of cleaner air and cleaner water far outweighed the costs (U.S. EPA 1997).

Here is the fundamental truth that it is time to face: just as we needed to develop new public health technologies to survive in cities with populations greater than one million, we must now invest in world-scale technologies to survive on a planet of seven billion people. The climate problem is the first planetwide stress we know about. Others will surely come. We need to learn how to develop and implement the twenty-first-century equivalent of indoor plumbing.

In addition to energy technologies, new technologies related to water, waste, and food should be developed. We need to develop the technology to reuse resources by applying energy to break them down to their component parts. An analogy is the way that a tree regrows its leaves every year but continuously grows its limbs. Water and soil, powered by the sun, are continually replenished by photosynthesis, resulting in new leaves. Cyclical growth is the basis for a renewable economy. We have to learn to use resources that regrow themselves every year, instead of relying on those that do not. The basic science and engineering of these processes are known. We need to improve the delivery and development of closed-system production processes throughout our economy. We are approaching the time in human, economic, social, and technological evolution when we have no choice but develop cost-effective sustainable technologies.

The Management Challenges of Sustainability

While it is easy to imagine that scientists and engineers can develop technological solutions to our problems, it is hard to believe that human organizations will be capable of adopting these technologies. Humans and their organizations are slow to change. Many of our current organizations are built to deliver goods and services based on nonsustainable technologies. The standard operating procedures that these organizations rely on are persistent and slow to change.

Nevertheless, technologies change and organizations can be transformed. Incentives are required to change behavior, and resources are required to pay for incentives. The benefits of new technologies can have a rapid and dramatic impact on organizational behavior and structures. Two decades ago, few organizations had information technology departments or chief information officers. Today, few large organizations do not have people performing these functions.

Sustainability requires new organizational capacities, organizational learning, and the education of individual organizational members. In part, people have to learn to think about resource use and waste in a new way. A study by MIT and the Boston Consulting Group (2009) asked seventy-five heads of global organizations and sustainability thinkers: "What will organizations need to be good at in order to thrive in the emerging sustainability economy?" Top answers included integrating sustainability into strategy, understanding integrated systems, collaborative innovation with stakeholders, and valuing long-term measurement and reporting (MIT Sloan Management Review 2009).

One aspect of the challenge is to develop more accurate measures of the immediate and projected impact of our actions on the environment. We also need to learn how to analyze impacts, develop methods for mitigating impacts, and communicating all of this to the senior managers who allocate resources.

The Political Challenges of Sustainability

Sustainability presents a series of political challenges. First, it requires long-term thinking. Our political process is oriented toward the present. There are people and interests who continue to believe that we must trade planetary sustainability for economic growth. We will continue to see a heated and probably symbolic debate between a "green side" and an "economic growth" side. On issues such as climate policy, we often hear scientists and environmentalists testifying before Congress that the approach is inadequate and too slow. Some business leaders and free-market advocates will say that an emphasis on sustainable development will ruin capitalism and the economy. I find neither argument persuasive. The economics of sustainability will not impair economic growth. In fact, environmental protection tends to fuel rather than impair economic growth. Just as previous environmental rules forced technological innovation, we found that environmental law tends to fuel economic growth (Shrivastava 1995). As for the argument by some environmentalists that we will not reduce climate change or preserve biodiversity well enough or fast enough—that is the fundamental question, to which no one knows the answer.

Scientists sometimes find politics frustrating, in part because of the difference between the scientific method and the policymaking process. Science tests hypotheses and builds mathematical models to gain knowledge and solve problems. Science is goal seeking and rational. The policy process is different. Policymakers don't actually try to solve problems; they try to make them less bad. The goal is not to solve the problem but to "move away from it." In New York City, we reduced homicides from more than two thousand a year to fewer than five hundred—the problem isn't as bad, but it is far from solved (New York State Division of Criminal Justice Services 2008). We aren't always capable of destroying the wild beast, but we somehow manage to keep it away from our door.

Policy, to quote the great public policy scholars David Braybrooke and Charles Lindblom (1963), is "remedial, serial, and exploratory." That means public policy tries to (1) remedy the worst parts of society's problems and (2) solve problems through trial and error. Most efforts to solve public policy problems are not a continuous process from start to finish. We start, we catch our breath, we reconsider—and then we start again. We make public policy this way because the problems that we ask governments to address are more complicated than the problems we assign to science. Environmental problems are caused by human interactions with our biosphere. Human beings and the biosphere are hard to understand. Add culture, economics, and technology to that mix, and you see why human social behavior is so difficult to predict. Even simpler behavioral questions, like "How do I motivate a teenager to clean her bedroom?" sometimes seem beyond our reach.

The punch line to the sustainability joke is that we will not solve it all at once. We simply do not know how to motivate all of the behaviors needed to solve the climate problem and other issues of global sustainability. In fact, we don't even know all of the actions that might allow us to solve the problem. At best, we are making educated guesses. We are in for a lot of "two steps forward and one step back." When you're in a crisis, as I believe we are, the key is to take those steps quickly. We also need to measure the results aggressively—even ruthlessly—and take corrective action when we make mistakes.

The politics of the sustainability policy process is wrapped up in symbolism as well as in the conflicts between narrow and community self-interest and between short-term and long-term self-interest as well. Politics, like the private sector, focuses on the next quarterly report or the next election. Success in these environments tends to be measured with little concern for the long term. The issue of sustainability is inherently long term. Successful pursuit of sustainability requires long-term leadership, which is extremely rare in politics. Even leaders sympathetic to sustainability goals will argue that they need to survive the short run to be around to address long-term issues.

The most successful examples of long-term leadership and action in government are probably in national security. There are also some examples of long-term thinking in infrastructure. But overall, long-term thinking is rare and difficult to accomplish. Unless we see a global ecological catastrophe, the politics of sustainability will always remain a challenge. If we have such a catastrophe, the challenges will be far worse than any presented by politics.

The Plan of This Book

As I indicated in the preface, after defining the field of sustainability management in chapter 1, the book will define and analyze the technical challenges, financial issues, management concerns, and the work of government in developing sustainable manufacturing and service businesses, renewable energy, sustainable water, a sustainable food supply, sustainable cities, and, overall, a sustainable planet.

My intent is to approach the issue of global sustainability as a problem of public policy and organizational management. What must be done to continue to grow our economy while preserving the planet's ability to generate wealth? In some cases, the answers are obvious, if difficult to implement. In other cases, we have no idea how to proceed. By the end of this book, the reader should understand the problem of global sustainability and have a rough road map of how we might begin to address the problem.

We start our progression toward an understanding of sustainability's importance with a look at the type of catastrophe that will only become more frequent if we do not work to reduce our reliance upon finite resources. Nothing expresses the urgency of implementing sustainability management as strongly as the 2010 BP oil spill in the Gulf of Mexico, and the case studies within this chapter, based on blog posts I wrote for the *Huffington Post*, serve as an introduction to the dangers of acting for the short term while ignoring the long term.

CASE STUDY: MANAGING SUSTAINABILITY IN THE GULF OF MEXICO

In May 2010, we learned that the U.S. Minerals Management Service issued permits to drill for oil in the Gulf of Mexico, despite warnings from the National Oceanographic and Atmospheric Administration (NOAA) that

(*continued*)

CASE STUDY: MANAGING SUSTAINABILITY IN THE GULF OF MEXICO *(continued)*

the drilling would likely harm the Gulf's fragile ecosystem. Of course, we can be grateful that Interior Secretary Ken Salazar made the decision to split the Minerals Management Service into three parts, separating staff that regulate safety from those that collect royalties for drilling on federal property. Unfortunately, all divisions will continue to report to the Department of the Interior, an agency infamous for selling natural resources to the highest bidder.

This is not a new story. The Obama administration is discovering failures throughout the world of federal regulation. Federal drinking water programs, air pollution programs, financial regulations, auto safety regulations, and of course, resource-extraction regulations are all in various states of disrepair. We saw evidence of this in New York City when the EPA was pressured by the Bush administration to give the "all clear" to reoccupy lower Manhattan immediately after 9/11. For too long, the regulation of business has been painted as illegitimate government intrusion in commerce.

Given the wealth they can deploy, it is not surprising that the business community and its lobbyists have managed to create an image of government regulation as un-American and vaguely socialist in origin. Imagine if we had the same attitude toward traffic lights on a busy intersection. ("Drive baby drive!") I realize that red lights restrict the freedom to drive, but don't we need red lights in order to have green lights? And what about those amber lights? What's that all about?

The problem is that population growth and the emerging global economy have increased the complexity and volume of economic transactions worldwide. This requires more rather than less regulation. However, in order for this regulation to be effective and to truly promote rather than destroy the economic wealth we seek to generate, regulation needs to be more sophisticated and better managed. Attacking science and cutting regulatory oversight is the wrong answer to the problem. So, too, is the symbolic battle between ecologists and industrialists. We need a realistic discussion of the risks we are undertaking to stoke the economic machine, and then we need a serious, well-managed effort to monitor and minimize those risks.

It is clear that the past decade and a half—the weakened Clinton administration and the antiregulatory Bush administration—has had a devastating effect on the regulatory capacity of the U.S. federal government. The attack on scientists, analysts, and decision makers has been reversed by the Obama administration, but as the Gulf spill clearly demonstrates, the effects of this antiregulation era will be with us for a long time. The almost cultlike glorification of the free market needs to end. There is no question that the capitalist market provides enormous social good and creates the wealth we all enjoy. But the rule of law is still needed. Free markets can't do everything. We don't want the mafia to run the trucking industry, and we don't want gangs to run amok in our neighborhoods. The profit motive can do a lot of good, but without the rule of law, it can also result in great harm. It is an indication of how idiotic this dialogue has gotten that I find it necessary to defend the very idea of regulation.

We need rules, but regulation, while necessary, is far from sufficient. We also need organizational capacity, and that requires resources. Cut the New York City police force from forty thousand to thirty thousand cops, and you will see the crime rate spike. Cut regulatory oversight and enforcement, and you will see oil rigs and coal mines explode, resulting in death and destruction. However, resources are not enough. When LeBron James was still with Cleveland, I remember watching him and his highly paid teammates lose to the Celtics. Cleveland had the capacity to win, but they needed leadership and motivation, which seemed lacking. Organizations are like sports teams. They can perform listlessly and go through the motions, or they can act with energy and enthusiasm.

Over the past quarter century, America's regulatory agencies have been attacked and defanged. This began with Reagan's notion that government was the problem, and it continues despite sporadic efforts to improve and reinvent government performance. At virtually every level of American government, the philosophy of "starve the beast" tends to dominate. Most of the public service and mission-driven students I have taught since the 1990s have headed toward nonprofits and away from government. Yet the policing function that we need if we are to manage the transactional and technological complexity of the modern economy requires government capacity. Government needs resources to attract talent and brainpower, but then it must apply those resources with energy, skill, and determination. Organizations are subject to a peculiar kind of gravity. If they are not moving forward with energy and drive, they quickly settle into mediocrity, sloppiness, and the type of performance we have seen in the incompetent regulation of oil drilling and coal mining. It's time to invest in that capacity or suffer the economic and ecological consequences.

MANAGING SUSTAINABILITY IN THE GULF OF MEXICO II

In the midst of the environmental disaster in the Gulf during the summer of 2010, the pundits began to focus on President Obama's management style and his lack of management experience. In an incisive *Politico* piece

(*continued*)

MANAGING SUSTAINABILITY IN THE GULF OF MEXICO II
(continued)

published on June 6, 2010, Glenn Thrush and Carol E. Lee observed that "the Gulf crisis has shed light on the strengths and weaknesses of Obama's unique management style, which relies on a combination of his own intellect, a small circle of trusted advisers and a larger group of outside experts. But it's also driven home a more generic lesson all presidents learn sooner or later: Administrations are defined, fairly or not, by their capacity to control stagnant backwater agencies, in Obama's case the Minerals Management Service, which failed to detect problems with the Deepwater Horizon well."

This, of course, is true, but this piece and others like it focus on the president and his approach to management and fail to discuss the far more critical issue of the now three-decade-long attack on the federal government's organizational capacity. Ronald Reagan began the process of dismantling the federal government's capacity. This effort to "starve the beast" and destroy federal capacity was reversed during the Clinton era, with Vice President Gore leading a well-intentioned effort to reinvent government, but the forces of disintegration regained momentum during the Bush years of 2001 through 2009.

During the presidency of George W. Bush, federal agencies that needed to build capacity for a new task were required to demonstrate that the capacity could not be found and purchased in the private sector. The underlying assumption of federal management during the Bush presidency was that government was the enemy and that the private sector was the great repository of management competence in America.

BP and the Gulf oil spill, Enron, the Wall Street meltdown, and the collapse of the American auto industry provide ample evidence that the private sector does not have a monopoly on management competence. Let's take a closer look at private-sector management. According to the American Bankruptcy Institute (reported in the Kansas City Business Journal on August 25, 2009), "more than 30,000 businesses filed for bankruptcy protection in the first half of 2009, up 64 percent from the nearly 18,500 in the same period last year." It is true that not all bankruptcies are caused by incompetence, and not all incompetence leads to bankruptcy. But it is also true that government organizations are capable of impressive accomplishments. Moreover, some of the work needed by our society—for example, inspecting oil rigs in the Gulf of Mexico—is best performed by government agencies. Unfortunately, the U.S. federal government has lost a great deal of its fundamental capacity over the past three decades.

This has happened because of a relentlessly ideological approach at the federal level to what management experts refer to as the "make-or-buy decision." This requires that every well-managed organization constantly ask itself: "Should we do this in-house or should we outsource?" That is a question that should be addressed pragmatically: "what would work best?" Management in the U.S. federal government has the answer provided for them: buying from the private sector is better than making it in the government. In an article I wrote in *Public Administration Review* in 2001, I argued for what I called "functional matching." I wrote that some tasks are best performed by government (especially policing), some by nonprofits (for example, mission-driven health and social welfare programs), and some by private firms (customer-driven services and manufacturing).

At the local level, government services are visible and have an immediate impact. While ideology plays a role locally, it doesn't usually dominate. In New York, the debate over charter schools has an ideological component, but the visibility of education performance measures provides evidence that moves the argument beyond ideology. Local officials are instantly accountable if water is not delivered, waste is not removed, fires are not put out, or criminals are not apprehended. In New York City, most social services are now delivered by nonprofit organizations under contract to the city's government. No one thinks about this practice as an ideological privatization strategy. It's simply the best way to help people in need. As a result of constant pressure to do more with less over the past three decades, New York City's government has improved its performance and capacity.

In Washington, D.C., symbolism and ideology drive agency management, and performance takes a back seat. The story at the federal level is characterized by management incompetence. We have seen it in the Department of the Interior during the Gulf oil spill, in FEMA's horror show during Katrina, and when we analyze the overuse of contractors and the overly small military presence during the second Iraq war. The lack of concern for capacity and management excellence has driven superb civil servants out of public service, destroyed government organizational capability, and made it impossible for the government to keep up with a more complicated and technologically based economy. The result has been the type of government performance we saw during the Gulf oil spill.

It would be helpful if the president had showed more leadership on the environmental catastrophe in the Gulf of Mexico. It is absolutely essential that he focus on the management of the organizations responsible for policing and protecting our environment, workplaces, and economy. However, the reconstruction of organizational capacity within the federal government will take many years, substantial resources, and incredible persistence. It will also require an ideological cease-fire that would return the "make-or-buy decision" to the purview of government managers. Can we do it? I seem to remember the answer to that question . . . oh, yeah: "yes we can."

Chapter 2

Sustainable Manufacturing and Service Businesses

What Is Sustainable Business?

The idea of sustainability is an outgrowth of the movement to protect the environment. The original notion of environmental protection involved regulating pollution created by manufacturing or transportation. Most solutions were "end-of-the-pipeline" add-ons. Pollution, while still produced, was treated with some type of technological filter that made it less dangerous. But retrofits add to the cost of production, and thus the popular belief developed that environmental protection makes goods more expensive. From there, it was only a short journey to the notion that there must be a trade-off between environmental protection and economic growth. Environmental protection and a concern for sustainability gained the reputation of being money-losing strategies.

If pollution is conceptualized as a form of waste, then the best-managed operation would be the one that produces the least amount of waste, not the one that "treats" and decontaminates the most waste. If the product is built with as few raw materials as possible, and if those materials are not scarce or finite, then the product is likely to be less expensive to make. This will not be true in every instance, but it will be true on average. Saving energy or developing renewable energy sources will undoubtedly make our economy more efficient in the long run; the problem is that our corporations and governments are managed on a quarterly or annual basis. In order to bring long-term perspectives into private organizations, we need to take explicit steps and invest public resources that will make it easier for the private sector to sacrifice short-term gains for long-term profits.

This is not an impossible task. The tax code has long been used to stimulate long-term investment. Investment tax credits and rules on depreciating capital investments are examples of these methods. The individual tax deduction for purchasing a home can be viewed as a method of facilitating long-term investments in housing. Despite the short-term focus of our organizations, a field of sustainability engineering is well underway, especially within the field of industrial ecology. The goal of sustainability engineering is to develop production processes that create as little waste as possible while depending on as few resources as possible.

The Technical Challenge of Industrial Ecology Defined

According to Nicholas Gallopoulos, industrial ecology can be defined as the

> discipline that considers industrial and commercial enterprises as an ecosystem analogous to biological ecosystems. Its organising principle is that industrial systems should emulate the best features of biological ecosystems, thereby reducing energy and material consumption and waste generation. The benefits of such operations are reduced environmental damage and increased sustainability for both natural resources and human activities.
>
> (Gallopoulos 2006, 10)

The concept of industrial ecology provides a framework for thinking about industrial production in a more sustainable way. All inputs of energy and materials are used in a closed system to create an output. While the concept of a closed production system works as a goal to aspire toward, its execution in reality is difficult. In some cases, the technology is not in place to reduce or reuse waste. There are two basic concepts related to industrial ecology: (1) waste prevention/source reduction and (2) reuse. According to the EPA:

> Waste prevention, also known as "source reduction," is the practice of designing, manufacturing, purchasing, or using materials (such as products and packaging) in ways that reduce the amount or toxicity of trash created. Reusing items is another way to stop waste at the source because it delays or avoids that item's entry in the waste collection and disposal system. . . . Source reduction, including reuse, can help reduce waste disposal and handling costs, because it avoids the costs of recycling, municipal composting, landfilling, and combustion. Source reduction also conserves resources and reduces pollution, including greenhouse gases that contribute to global warming.
>
> (U.S. EPA 2010e)

In addition to using fewer materials and avoiding waste in the production process, industrial ecology builds on the concept of reuse. The idea is that even when a product comes to the end of its useful life, we think about the ways that it can be remanufactured or repurposed. Anita Ledbetter (2009) discusses reuse in the context of "green building":

> Reuse is a primary concept in green building practice. The total environmental cost of incorporating this concept along with other green building strategies, in whole considered as "lifecycle cost" helps to determine the depth of environmental impact. . . . By incorporating the concept of reuse into the design of a building, an architect considers not only construction materials, but also site selection, landscaping, mechanical and electrical systems, energy exchange, and minimizing waste in the construction process.

Green building is one of many industries incorporating sustainable practices into daily operations. In the business world, there is a growing awareness of the meaning of sustainable business and an increasing understanding of the importance of environmental constraints and opportunities in manufacturing. Thanks to this awareness, a number of companies are implementing environmental management systems, defined by the EPA as "a set of processes and practices enabling an organization to reduce its environmental impacts and increase its operating efficiency" (U.S. EPA 2009). Environmental management systems are based on the Porter hypothesis, which claims that tightening environmental standards increases a firm's profits by eliminating unnecessary, expensive inputs. In moving from a compliance-based, "end-of-pipe" approach to a preventative approach emphasizing source reduction and process innovation, environmental sustainability can actually increase, not reduce, a firm's profits (Russo and Fouts 1997). The International Organization for Standardization is working to help corporations create preventative environmental management systems, with its ISO 14000 management standards. The ISO 14000 certification family provides guidelines and requirements for companies' environmental management systems. The International Organization for Standardization's handbook from 2004 explains:

> Unless a structured approach is taken, the organization may focus on what it believes to be its environmental impacts, a belief based upon "gut feel" and ease of implementation. In reality, this does not address real issues but promotes a "green" feel-good factor or perceived enhancement of image—both internal and external to the organization—which is not justified. For example, a company engaged in the extraction of raw materials by mining may have an environmental objective to . . . "save energy by switching off lights." There will be some energy saved by administration personnel switching off lights and heating when they are not being used for long periods. However, such savings in energy are trivial compared to the massive impact that the mining industry has on the environment.
>
> (Whitelaw 2004)

The ISO regulates and encourages the sustainability practices of firms interested in certification. Once a company achieves ISO 14000 certification, it must continue to meet future sustainability goals. While the public image that certification gives a firm is a financial incentive for companies to become more sustainable, many companies have found sustainability management itself to be financially beneficial. For example, the coffin manufacturer J. C. Atkinson & Sons saves approximately $27,000 annually in gas and oil costs by heating its factory with waste wood scraps. In a 2009 interview, Amory Lovins points out "a very large beverage maker that could, we found together, make more money from something it was paying to throw away than from its main product" (Hopkins 2009, 37). The idea is that if firms can develop good enough management practices, they can tighten their production processes, minimize useless waste, and make or save money. This shows that in certain cases, the trade-off between sustainability and financial success is only a misconception.

In a 2006 survey of sustainable manufacturing policies, Hartmut Kaebernick and Sami Kara (2006, 28) concluded that

> the majority of companies acknowledge the importance of environmental requirements but they follow different paths of implementation with different priorities. It is an agreed perception that environmental issues will become more important in the near future. Product development activities already widely include environmental requirements with a range of design tools mainly used in Europe and the USA. The effects of environmental product development have led to higher costs and better product quality. Product recovery is widely considered as an option, but not many companies are directly involved in the collection process.

There are a number of business opportunities and cost savings that are created by the move toward industrial ecology and environmental management systems. An example of a business that has benefitted from these initiatives is the California Materials Exchange (CalMAX), which is built on the idea that "one business's trash is another business's treasure":

> Businesses, schools, and nonprofits can utilize CalMAX to search for available and wanted materials. . . . CalMAX conserves energy, resources, and landfill space by helping businesses and organizations find alternatives to the disposal of valuable materials or wastes through waste exchange. CalMAX supports educators by offering free materials that support the implementation of curriculum

> instruction or art projects! Businesses are encouraged to support their local schools by listing any free overruns, overstock items, and reusable materials on CalMAX.
> (California Department of Resources Recycling and Recovery 2010)

The CalMAX program is an example of the financial opportunities that can be created by including waste in our definition of what a resource is. In one case, a city government's trash becomes a design firm's resource:

> One successful exchange occurred between Gia Giasullo, a partner of the Oakland-based design firm Studio . . . and the city of Morgan Hill. Giasullo acquired blueprint paper—a material difficult to recycle because of its ammonia content—from the city of Morgan Hill and used it to print her firm's brochure for office furniture. She was so pleased with the brochure's appearance that she created a new line of consumer stationery called "Used Blues" from the discarded blueprints of architects and engineers.
> (ACFNewsource 2002)

By encouraging recycling, a governmental organization encourages private business to become increasingly environmentally sustainable and at the same time creates an opportunity for innovation and success. In 2002, CalMAX estimated that its program saved participants over $3 million and diverts around 600,000 tons of materials from going into landfills (ACFNewsource 2002).

The Challenge

In some cases, the technological base for reuse or waste reduction has not been developed. Energy is often required to reuse products. Many manufacturing technologies have been built on the expectation that waste disposal was less expensive than reuse. While reduced resource use is often more obviously cost effective, many technical standards require overbuilding and manufacturing in excess of standard as part of the definition of quality. An engineer may require that a steel rod be five times as thick as needed—just to make sure it never breaks.

In many respects, the technical challenge is not particularly profound. It is possible to design an easy method of reusing a product. In 2005, Eriko Saijo of Japan for Sustainability presented an excellent case study of Fuji Xerox Company's effort to redesign the office copier. Fuji Xerox's term for green manufacturing is what they call "inverse manufacturing." According to Saijo (2005), this is:

> A manufacturing system that considers the entire product life (planning and designing, manufacturing, use and disposal) at the design phase and incorporates into the design consideration of the process of collecting used products, dismantling, separating, and reusing them in the form of components and materials. . . .
>
> Inverse manufacturing minimizes the consumption of resources, energy and the volume of waste, while also creating value. . . . Inverse manufacturing has been propounded as a model for sustainable manufacturing over the past dozen years or so, but practicing it is not easy and only a limited number of companies have succeeded in adopting it as a business model.

Since 1995, Fuji Xerox has made copy machines that are designed to reuse components. They try to make components that can be used in a variety of copy machines. When designing new machines, they begin by trying to maximize the use of parts from previous models. As Saijo (2005) observes:

> The most remarkable feature of Fuji Xerox's inverse manufacturing system is to enhance reusing of parts not just for the current product models but also for future models, up to three product generations hence. The parts designed only for one model cannot be reused when the applicable model is discontinued, even if technically they are not end-of-life yet. However, if the parts are designed as reusable for different models, they can be applied even for new models and thus continue to contribute to reducing environmental impacts as a result. In fiscal 2003, about 60 percent of reused parts were applied for third-generation models.
>
> With such efforts, the company succeeded in increasing the reuse ratio of components to 54 percent in fiscal 2003. Also, in the same year, new natural resource input was decreased by 2,200 tons

a year, which accounts for about 3.5 percent of the total ar
resource input. This means that Fuji Xerox made it possil
duce three out of a hundred copy machines without using any ne..
natural resources. Furthermore, the company was successful in achieving zero waste by sorting out nonreusable parts thoroughly to recycle them.

Xerox has also been a sustainability leader in the United States for about two decades. According to Jack Azar, Xerox's vice president of environment, health, and safety:

> Xerox adopted a strong environmental policy in 1991. This policy, along with several early waste-free successes, firmly established these principles in the company. By success, I mean that we not only substantially reduced waste, but also we realized substantial cost savings. We describe our commitment as seeking to produce waste-free products in waste-free factories in order to enable our customers to be waste-free.
>
> Xerox products are designed to reduce environmental impacts in all phases of the life cycle. We design our products for remanufacturing and design imaging supply items for return, reuse and recycling. These early decisions mean that once the product's supply item or the product itself has reached its end-of-life, we maximize the assets and minimize the waste. These programs divert over 150 million pounds of electronic waste from landfills each year. Since 1991, over 1.5 billion pounds of product waste have been reused or recycled. We also design our products to be energy-efficient and to make efficient use of paper. Both of these initiatives reduce waste during the product's use phase.
>
> (Baue 2004)

When reuse and reduced resource consumption become design specifications, the real work of sustainable manufacturing can begin. Just as changes in occupational safety practices have been introduced over the past century, we can expect these factors to gradually influence manufacturing technologies and processes in the future. The chief technical challenge will be to accelerate this process. Engineers, given the time and resources to do this work, will be able to achieve the goals of green manufacturing. A key issue then is the access to financial resources needed to fund the development and use of new technologies.

Financial Issues

Manufacturing processes and service delivery systems that use fewer resources and produce less waste in theory should operate at a lower cost than those that waste resources and must pay the costs of waste disposal. The reality is that sunk costs and the capital requirements of new technology have a negative effect on the payoff of sustainable business practices. Consumer preferences must also be factored into the equation. For some consumers, a piece of equipment made of "recycled" parts might be seen as lower quality than a product with all-new components.

The longer-term financial benefits of sustainable business practices are easy to analyze: lower energy costs, less funding allocated for raw materials, lower liability due to the reduced use and disposal of toxics, and so on. However, these are long-term benefits, and the world of business and its reward system is organized around short-term revenues, expenses, and profits.

Capital costs must be analyzed in light of their return on equity. Sustainability may generate financial benefits, but capital flows to the highest rate of return, and sustainability investments may pay off, but not at the level of profit needed. An additional factor that must be assessed is the relative scarcity of capital and its price. If capital is scarce, sustainability investments may not be attractive enough to compete for investment funds. If capital is plentiful, long-term sustainability investments will do better at attracting capital.

Another financial factor is the uncertain investment return related to the use of new technology. While some sustainability investments rely on off-the-shelf technologies such as double-paned windows and similar energy efficiency measures, others require technologies that are less proven. This increases the level of risk involved in these investments and can discourage companies from allocating capital to sustainability initiatives. In some cases, companies will wait to see if others in their industry take the plunge; in other cases, they will simply reject the investment as too speculative.

While the capital requirements for sustainable business practices can stall efforts to adopt these practices, there are some promising trends in the "green marketplace." A 2009 study of green jobs by the Pew Charitable Trusts indicated that the Obama administration's early focus on sustainability as an economic recovery strategy was sound economics. This government initiative appeared to be following rather than leading trends in the marketplace, and the government may very well have managed to throw its money in the right direction during the 2008–2009 recession. According to the Pew study:

> The number of jobs in America's emerging clean energy economy grew nearly two and a half times faster than overall jobs between 1998 and 2007. . . . Pew found that jobs in the clean energy economy grew at a national rate of 9.1 percent, while traditional jobs grew by only 3.7 percent between 1998 and 2007. There was a similar pattern at the state level, where job growth in the clean energy economy outperformed overall job growth in 38 states and the District of Columbia during the same period. The report also found that this promising sector was poised to expand significantly, driven by increasing consumer demand, venture capital infusions, and federal and state policy reforms.
>
> (Pew Charitable Trusts 2009, 4–5)

The Pew study carefully defined green jobs to include employment in: "(1) Clean Energy, (2) Energy Efficiency, (3) Environmentally Friendly Production, (4) Conservation and Pollution Mitigation, and (5) Training and Support" (Pew Charitable Trusts 2009, 5). This was an important and methodologically sound study that both defined green business and measured its early growth. There are increasing signs of the mainstreaming of green business as it has moved out of public relations and "greenwashing" into the world of hard-headed, realistic business practice.

Despite the financial issues cited above, there is no question that this field is growing. These financial challenges will remain, but ultimately they will be seen as routine rather than as special cases of business capital requirements. Investment capital is always an issue in any business. The trend that I see is that the capital requirements for sustainability management will come to be seen as simply a part of the capital needed for business modernization.

Organizational and Management Challenges

While capital is a necessary condition for more sustainable business operations, it is by no means a sufficient condition. Sustainability in many organizations requires a dramatic change in management culture. An organization with an aggressive, "damn-the-torpedoes" approach to work processes and management must typically learn to slow down and think. Organizations that think of wasted water, energy, and resource use as "breakage" or simply "the cost of doing business" often require a complete cultural transformation.

Large organizations often develop the equivalent of interest groups that push to maintain a particular set of practices because their status and clout within the organization are derived from those practices. For example, during 2009, as General Motors neared bankruptcy, these vested interests made it nearly impossible for the company to recast itself as a modern auto company. The ingrained belief that "mini cars" resulted in "mini profits" was an unshakable part of GM's corporate culture. In the end, much of the senior management of the company had to be replaced in order to begin the process of rebuilding the company (Muller 2009). This is, of course, not limited to the private sector. Large-scale government agencies, from local public-housing agencies to the U.S. Department of Defense, have the same problem.

Part of the issue actually stems from the engineering mindset of overbuilding for the sake of safety and security. For maintenance staff operating a building, the repercussions of breakdowns are far greater than those of wasting resources. The need for an adequate margin of error must be understood and factored into any effort to influence this type of thinking. Similar biases generate waste in many other types of organizations. Of course, some waste is simply a result of sloppiness, incompetence, and inadequate management. It is not overbuilding that is the problem, just laziness and idiocy.

In addition to these almost ideological (or at least cultural) issues, organizations also face the typical challenges involved in any form of large-scale organizational change. New standard operating procedures must be developed and new technologies learned and adopted. Staff must be trained to understand how to perform new tasks and how to use new equipment. Management must learn enough about the new way of working to supervise these tasks, and, of course, this must take place without sacrificing significant levels of production during any period of transition.

The changing technological base of modern production and the impact of rapid communication along with the global economy have put a range of pressures on many organizations. Sustainability is not the only challenge faced by modern businesses and government organizations. In some cases, this has resulted in organizations that have become more agile and better able to manage change. In other cases, we see companies such as General Motors resisting change until they are forced into bankruptcy. The need to adopt green business practices can have the effect of helping an organization become better at managing change. To paraphrase Friedrich Nietzsche: what doesn't kill you makes you stronger.

In large organizations, another management challenge of sustainable business is the need to communicate new policies, priorities, and procedures throughout the organization and to coordinate actions requiring multiple organizational units in multiple locations. Fortunately, the technology of information management and organizational communication has become more advanced and less expensive. This can facilitate rapid change and help ensure that organizations are able to learn new ways of working faster than ever before. In the end, of course, it all comes down to the human ability to adapt, learn, and change behavior. These are not easy things to do, but they are, nevertheless, done all the time.

If this sounds theoretical and pie-in-the-sky, it's not. In a study reported in June 2009, the Aberdeen Group benchmarked over two hundred businesses and found that "best-in-class companies reduced their energy costs by 6 percent, compared to an industry average of a 4 percent increase. These companies also reduced their carbon footprint by 9 percent and paper costs by 10 percent. They also cut both facility and transportation/logistics costs by 7 percent." Overall, they found that sustainability initiatives resulted in a cost savings ranging from 6 to 10 percent (Environmental Leader 2009).

It is true that business practices change slowly, and even potential cost savings and support for the principles of green business will not always result in changes in business practice. In a very interesting 2008 study of 1,600 logistics executives, the outsourcing experts Capgemini along with the Georgia Institute of Technology, Oracle, and DHL found that while "98 percent of respondents said green supply chain initiatives are essential for future business success . . . the majority are unwilling to invest additional funds in green supply chain" (GIT 2008). According to Dr. John Langley of the Georgia Institute of Technology: "The greatest shared challenge is that of forming and growing successful collaborative relationships between users and providers of logistics services. Without a commitment from both sides little progress can be made in the greening of the supply chain and supply chain security" (GIT 2008).

The Role of Government and Public Policy

When one looks at how private organizations come to change their practices, competition is often the single most important factor, but in addition to market forces, one sees the key influence of public policy and government regulation. Investment decisions and mass consumer behavior are both influenced by the tax code. The deductibility of mortgage interest and home property taxes along with government-insured mortgages helped transform the United States from a nation of renters to a nation of homeowners (Lowenstein 2006). A tax on gasoline, a tax credit for renewable energy, and other forms of tax expenditures can be powerful tools to promote sustainability. In addition, regulations on greenhouse gases, coupled with some type of price on carbon emissions, will do a great deal to increase the consumption of renewable energy.

In addition to the use of regulation and tax policy, government has a critical role to play in funding and managing research leading to the development of new sustainability technologies. The private sector also has an important role to play in developing new technologies, but private efforts cannot substitute for government-funded research. One can imagine a continuum from basic scientific research on one end to applied technology development on the other. Privately funded research tends to be more applied, and government-funded research tends to be more basic. Without the basic research, there is no new knowledge to apply, and thus government's role is critical. There is a time in the development of a new technology when the potential for a commercial product is too far in the future and too uncertain to generate private investment capital. That is when we need to rely on government.

Finally, government can play a role by providing funding for modernization directly to businesses. California's program to promote energy efficiency is an excellent example of a creative partnership between government and the private sector. California's program began in 1996 and required the three major private utilities to collect $540 million annually in fees to promote "public purpose" programs, with $220 million set aside specifically for energy efficiency. In 2002, Californians began to pay another smaller surcharge to promote renewable energy. The renewable energy fee generates $135 million annually. The other goals of California's energy efficiency program are to improve the efficiency of appliances sold in California and to increase efficiency standards for buildings (CPUC 2009). In September 2005, the California Public Utilities Commission (CPUC) authorized $2 billion in energy efficiency funding for 2006–2008, the most ambitious energy efficiency and conservation campaign in the history of the utility industry in the United States. According to the California Energy Commission, "California's building and appliance standards have saved consumers more than $56 billion in electricity and natural gas costs since 1978 and averted building 15 large power plants. It is estimated the current standards will save an additional $23 billion by 2013" (Garfield et al. 2007, 3).

California is the most energy-efficient state in the United States. While per capita electricity consumption in the United States increased by nearly 50 percent over the past thirty years, California's per capita electricity use remained almost flat, due in large part to cost-effective building and appliance efficiency standards and other energy efficiency programs (Garfield et al. 2007, 3). California was able to achieve this efficiency by establishing building codes that require energy efficiency in new construction and by stimulating private companies to improve the efficiency of their own operations. The way they have made this happen is very clever. Instead of growing a government agency to work on energy efficiency with the private sector, the funds collected for energy efficiency in California are allocated to the state's private electric utilities. However, the utilities can only spend the money on energy efficiency projects. Since utilities are in the business of generating and distributing energy, not saving it, most of the energy efficiency work in California has been outsourced to companies that specialize in that type of work. In fact, California's market for energy efficiency has given rise to around forty businesses that focus exclusively on building and enhancing energy efficiency.

The creation of energy efficiency service providers has created an industry of firms that make their money by making other private organizations more energy efficient. This creative mix of private gains with public aims works, and it is a model for an effective sustainability policy. However, we have a great deal to learn about how to promote sustainable business practices and the effectiveness of those practices. We currently are seeing the start of efforts to develop metrics for measuring these initiatives, and, on the part of the U.S. government, the development of programs to assist firms. My own belief is that government should not attempt to get into the business of sustainability consulting. While the model of agricultural extension may be tempting, we are a long way from the period when land grant colleges developed the technology that allowed for agribusiness and the Agricultural Extension Service taught farmers how to use new methods and technologies. Instead, public policy should focus on providing incentives and resources, and the government should provide clearinghouse, coordination, and information services to support private efforts.

An important issue for government to focus on is the definition of sustainable business practice. The European Community recently completed a study of this, and a number of think tanks are focused on definitional issues as well (OECD 2009). The problem is the tendency to confuse greenwashing and public relations with substantive, measurable, sustainable business practices. An increasing number of businesses, such as Wal-Mart, are serious about sustainability and work to integrate it into their daily operations. Others, however, want to appear "green" but could care less about resource use or pollution control. Government and nonprofit organizations can play a role by defining sustainability and possibly providing certification to organizations adhering to these practices. In the case of the energy efficiency of buildings, this certification has been effectively performed by LEED. While LEED is not a government agency, it is administered by the U.S. Green Building Council, a nonprofit organization with a deep commitment to sustainability principles that plays an important role in certifying sustainability.

Both nonprofit organizations and the government have a role to play in certifying sustainability, although only government has the authority to regulate the certification process. In general, the role of government and public policy should be to selectively and creatively motivate and police private and nonprofit efforts to develop and manage sustainability practices. Government must carefully calibrate its function so as to play a role that can't or won't be played by private parties. When private organizations take on these roles, government should withdraw and move on to new ones.

In order to gain a real-world understanding of sustainability management, it is important to take a look at companies that have successfully implemented it. Xerox has demonstrated its commitment to the environment and been financially rewarded for its judicious resource use.

CASE STUDY: SUSTAINABILITY MANAGEMENT

Q&A with Patty Calkins, vice president of environment health and safety at Xerox. E-mail interview in July 2009 by the staff of the Earth Institute, at Columbia University.

Q1. Xerox has blazed a trail in terms of its commitment to environmental issues and questions of sustainability. What advantages has this advance position reaped for the company?

Our commitment to sustainability benefits our business and our bottom line. Long before going green was popular and sustainability entered our daily lexicon, Xerox embedded environment sustainability practices and principles into all aspects of our business and throughout our supply chain. As a result, our strategy and the programs we've implemented have proved through financial results that sustainable initiatives are good for the bottom line. They have also established that innovative and sustainable design, engineering, manufacturing, and marketing produce environmentally sound products and services that can reduce our customers' environmental footprint as well as our own.

Here's what we've done at Xerox:

1. Xerox invented toner: the ink that puts the mark on the page. EA (emulsion aggregation) toner is a chemically produced toner—a Xerox-patented approach—that is created using 60 to 70 percent less energy per page than conventional toner. Unlike traditional toner, which is created by physically grinding materials, EA toner is chemically grown, enabling the size, shape, and structure of the particles to be precisely controlled, which leads to improved print quality, less toner usage, less toner waste, and less energy required for manufacturing and for printing.

(*continued*)

CASE STUDY: SUSTAINABILITY MANAGEMENT (*continued*)

2. In September, Xerox opened its doors to a new EA toner plant in Webster, New York, which is Xerox's most energy-efficient building to date. The five-story building has more than three thousand sensors that feed information about temperature, humidity, airflow, and other variables into a networked system. Depending on what's happening in the plant, entire zones of the building may be shut off to reduce energy use.

3. Xerox has long been focused on energy efficiency and reducing greenhouse gases as part of the company's commitment to sustainability. In December 2007, Xerox announced that it had achieved an 18 percent reduction in greenhouse gas emissions since 2002, exceeding its original goal of 10 percent. The company's new goal is a 25 percent reduction by 2012. To meet the target, we launched a company-wide energy-reduction program called "Energy Challenge 2012." The projects implemented resulted in significant greenhouse gas reductions and also helped Xerox save money—$18 million in 2006 and 2007. For 2008, we are estimating a savings of over $14 million.

Many customers have their own environmental and sustainability goals and expect their business partners to share a commitment to sustainability. Xerox delivers technology, products, and solutions to help our customers reach their environmental goals and the needs of their business. Here's proof:

1. Customers want to reduce energy use for cost and environmental savings. Our office services and Energy Star offerings can do that. In fact, customers can reduce energy use from their document technology by up to 50 percent when they replace single-function devices with Xerox multifunction devices.

2. Customers want to reduce their use of excess paper and use environmentally preferable papers. Xerox's digital technology reduces paper use through easy-to-use features such as two-sided printing and duplexing and electronic document management. We offer recycled-content papers and environmentally certified papers from the Forest Stewardship Council (FSC) and the Programme for the Endorsement of Forest Certification (PEFC).

3. Proof that a shared focus wins business: a team from Mississauga, Canada, earned Forest Stewardship Council (FSC) Chain of Custody Certification for onsite print services sites across Canada. As a result, several accounts cited FSC certification as a differentiator in their decision-making process. The value of these contracts totaled $12.5M.

Q2. How difficult was it for the company to implement concepts of reuse and reduced resource consumption (end-of-life equipment take-back, inverse manufacturing) in terms of convincing shareholders, corporate culture, cost, technol-

ogy, and so on? Did the approach begin at the top or the bottom, and how was it successfully manifested throughout the entire company? Did the management have to be changed at all?

Social responsibility has been ingrained in the Xerox culture since our earliest days as a company, in part because our founder, Joe Wilson, and the inventor of xerography, Chester Carlson, were such advocates of corporate citizenship. Xerox people have always shared that vision, and we've been a pioneer of environmental programs and processes. A few examples:

1. Dating back to the 1960s, Xerox introduced the first copiers with energy-saving standards and began recovering metals in photoreceptor drums for reuse and recycling.

2. In the 1970s, we invented "duplexing," or copying on both sides of the page.

3. In the 1980s, we introduced automatic "power-down mode" for our technology.

4. In the early 1990s, Xerox pioneered the practice of converting end-of-life electronic equipment into products and parts that contain reused parts while meeting new product specifications for quality and performance. We have developed a comprehensive process for taking back end-of-life products and have established a remanufacture, parts reuse and recycling program that fully supports our waste-free initiatives.

Xerox continues to improve upon our environmental sustainability programs and efforts for the benefit of the environment, our business, and our stakeholders. In fact, according to a Thomson Reuters study, 82 percent of investors evaluate environmental, social, and governance (ESG) criteria as part of their investment decisions, because they believe that ESG actions impact share price.

Q3. How did you overcome the question of uncertain returns on investment? What pushed you to take these trailblazing steps? Were short-term profits (as opposed to long-term profits) affected at first? How long did it take for the company to reap significant rewards?

Good citizenship and good business are not only compatible but synergistic. What started as the right thing to do quickly evolved into the right thing for our business and our customers. Waste is the result of an inefficient process. If you remove the excess waste, you lower cost and boost productivity at the same time—a win for the bottom line and the environment. Xerox's experience with reuse, recycling, and remanufacturing has not only kept waste out of landfills but saved the company more than $2 billion as it did so. We've also saved millions through our efforts to reduce energy use and reduce greenhouse gas emissions. And we've passed this on to customers through sustainable technologies and services.

(*continued*)

CASE STUDY: SUSTAINABILITY MANAGEMENT *(continued)*

Q4. Have you seen a big impact on consumer preferences? Were consumers difficult to convince in the beginning (the idea that recycled products might be of lesser quality, for instance) or were they a driving force?

There is proof that when combined with sound management and technology, a commitment to sustainability allows companies to get more out of existing investments. Customers look to us to help them boost productivity, with a laser focus on cost and quality. At a time when most companies are looking for ways to do more with less, an investment in sustainability strategies pay off.

Here are a few examples of how our sustainable approach to technology and services yields greater productivity and cost savings for customers, convincing them that what's good for the environment is also good for business:

1. Xerox developed and launched the industry's first Sustainability Calculator as part of Xerox Office Services. The tool evaluates the current office environment of printers, copiers, and multifunction devices and then measures environmental benefits that could be achieved in terms of energy and paper use, solid waste, water, air, and greenhouse gas emissions. For one sector of a global defense and technology company, the calculator was able to show life-cycle savings of 27 percent in energy usage, 26 percent in greenhouse gas emissions, and 33 percent in solid waste. This sector had more than two thousand printers and hundreds of multifunction printers and stand-alone copiers prior to leveraging Xerox Office Services—which optimized the office to fewer than 1,100 devices.

2. We did a paper audit of one of our large industrial clients. We found that it had 16,000 printers across its offices, producing 480 million printouts a year. While each document printed cost little, all that printing added up to $100 million over five years—and a huge carbon footprint. With just a little bit of hard work and by applying some of these tips, the firm managed to cut its number of printers to five thousand and now spends between 20 percent and 30 percent less on printing each year, with a much smaller environmental impact.

Q5. What are the main challenges for a company undertaking sustainable practices? Are the greatest challenges financial, risk related, technological, or cultural?

The world today is facing serious environmental challenges, and global solutions are being sought. At the same time, businesses are expected to be a part of the solutions amid increasing expectations of a multitude of stakeholders. When starting on the path to a more sustainable future, we share with our customers the following critical steps to success:

First, know where you started and measure your success—take a quantitative approach.

Second, any environmental or sustainability strategy has to map back to your business—it's great if you're green, but if you don't make money you won't be in business for very long.

Last, quick payback and easy wins are essential to gaining buy in from stakeholders, especially employees, and it establishes momentum.

The fact is that we cannot afford to be complacent. At Xerox, we feel as though our job is never done—to us, it is a journey that requires leadership, clear direction, and innovation that pushes through to the next frontier of what is possible.

Q6. What are the overall advantages for Xerox? Have sustainability practices promoted profits in the short and/or long term? Is it quantifiable?

The answer is absolutely, yes, and there are examples and anecdotes sprinkled throughout our answers as proof points.

Chapter 3

The Centrality of Energy

The heart of the sustainability issue is energy. Unless we end our dependence on fossil fuels, we cannot develop a long-term sustainable economy. There are three reasons for this: (1) Fossil fuels emit carbon dioxide and other pollutants when they are used. (2) Extracting these fuels damages the Earth's ecosystems. (3) These fuels are finite, and as they become scarcer and more difficult to extract, they will become more expensive. If we can develop low-cost renewable energy, we can overcome these constraints and go a long way toward addressing the current sustainability crisis.

As Howard and Elizabeth Odum first discussed in their classic, *The Energy Basis for Man and Nature*, all of the planet's systems and resources were created by chemical transformations caused initially by solar energy.

> Faced with shortages of energy, serious inflation, overgrowth, and concern for protecting the environment, human beings are coming to realize that they may be forced to change their patterns of life. It is now clear that our future depends on the connection of energy, economics, and the environment (sometimes called the three E's) into one system of interdependent actions.
>
> Both public policy and individual attitudes reflect a basic confusion concerning the relationships of energy, economics, and environment and their influence on the drastic changes taking place

> in our lives. . . . Fundamental questions are being raised about the future. . . . To understand these issues . . . we must understand the principles of energy, economics, and environment. . . . Simple energy-flow diagrams allow us to see the behavior of whole systems in terms of the interactions of resources and processes.
>
> In looking to the future, we have considered the . . . state of slowly declining energy. We see little evidence that any existing or proposed energy sources will yield sufficient net energy to cause major growth after the rich fossil fuels are gone.
>
> (Odum and Odum 1981, 13, 23, 260)

Modern life is built on a series of inventions that require energy—lighting, radio, TV, refrigeration, climate control, motor vehicles, and so on. What is less well recognized is that our system for growing and delivering food and for distributing water also requires massive amounts of energy. Energy from the sun fuels the process of photosynthesis, which all life depends on. Even fossil fuels originated as solar energy in the prehistoric past. But fossil fuels are created in geologic time, and for all practical purposes, once we use them, they will never return. Michael Common and Sigird Stagl have very succinctly described the relationship of fossil fuels to solar energy:

> Given the standard account of the fossil fuels, we can think of solar energy as being like a flow of money coming in as income, and then the fossil fuels are like a savings account into which deposits were made (by means of photosynthesis), a long time ago, from that income. The fossil fuels are saved-up past receipts of solar energy, where the saving was made possible because some solar energy was converted to plant tissue by the process of photosynthesis.
>
> (Common 2005, 43)

Or as Odum and Odum (1981, 115) conclude: "directly or indirectly, all the great systems of the earth run on solar energy." The use of technology to harness energy is what allowed us to grow enough surplus food to put off the decline in population projected by Malthus hundreds of years ago. All of this is to say: if we solve the energy problem, we solve the sustainability problem. Renewable, solar-based power is the transformative technological fix that we need to focus on.

Alternatively, a safer, smaller-scale, cleaner, waste-free form of nuclear power could also work. What we need is a power source that is low cost, less capital intensive, decentralized, and less polluting. Unless we can do a better job of harnessing the energy released by the atom, the sun is probably our best bet. That particular atomic reaction has already been provided for us.

The Technological Challenge

The development of our energy system proceeded in a haphazard manner during the twentieth century. In major metropolitan areas, private companies were granted franchises to operate public utilities that were local monopolies. These companies built huge central power plants and then built power lines to connect residents and businesses to the power. In some areas, natural gas lines were built the same way. Some natural gas was also distributed to households in containers transported by motor vehicles. Oil to fuel cars and heat buildings was distributed in tankers, which do not require rights of way for construction, as do pipelines. However, large corporations were needed to generate the capital required to discover, pump, and distribute oil. The systems were large scale and centralized, as one would expect in the industrial era. In rural areas, government had to intervene with projects such as the Tennessee Valley Authority, which built hydroelectric dams and other technologies to bring about rural electrification.

With the exception of some hydroelectric and nuclear power, most of our electricity is based on fossil fuels. The system is highly centralized, and a great deal of the electricity is lost between the place it is generated and the place it is used. The process of transmitting and using energy can be quite inefficient. This is closely linked to the three great technological challenges that sit at the center of the sustainable energy problem:

1. How do we develop renewable energy sources that are clean and cost effective when compared to fossil fuels? Can we develop a small, low-cost solar cell and a battery for storing electricity once we generate it? Can we develop a waste-free and less capital-intensive form of nuclear power?

2. How do we make the transmission process "smarter," more efficient, and open to decentralized transmission?
3. Can we learn to collect and store carbon dioxide from the fossil fuels we will continue to burn?

These technological challenges are solvable and in all likelihood will be met over the next several decades. My confidence is based on the convergence of economic and environmental factors related to sustainable energy. Without renewable energy, the price of energy as a proportion of the gross domestic product will probably rise. The impacts of fossil fuels on ecosystems and on global climate are well known. Just as computers have become exponentially more powerful and smaller, solar receptors and batteries are also capable of rapid technological advancement. Some form of nuclear power might also result in low-cost energy, although our initial experiences with nuclear have not been encouraging. While I do not dismiss the possibility of a nuclear breakthrough, I think that solar holds more immediate potential. I confess that this is a political and managerial analysis, not a technical one.

The computers of the 1960s became the building block for today's laptops. I expect to see the same pace of development around solar technology, because the motivation is so clearly in place. Governmental and private resources will be invested in developing these technologies, and once solar becomes cost competitive with fossil fuels, it will develop increased momentum and ever lower prices.

The issue of energy transmission has also gained increased attention, with the development of what has been termed a "smart grid." According to Tom Casey, the CEO of Current Group LLC—a smart-grid software and analysis company: "A smart grid in many ways is like an Internet for electricity, a network of devices that are monitored and managed with real-time communications and computer intelligence" (Ling 2009).

The system that transmits electricity is called the electrical grid. In the United States, the grid has evolved piece by piece over the past century, but much of its technology is out of date. An electricity grid is not a single piece of infrastructure but a collection of distinct distribution networks and power generation companies. Smart grids use computer controls to increase the coordination between parts of the electricity transmission network. If we are going to move to a system that is powered by the decentralized generation of renewable energy, our grid must be updated to receive and transmit this energy.

The final technical challenge is that of energy efficiency. To some degree, this is not totally an issue of technology but one of institutional inertia and the very human tendency to resist change. In the United States, we use and waste enormous amounts of energy. With the price of energy relatively low, there has been little incentive to design cars, homes, and appliances that use less energy. While we will turn to the organizational issues below, let's first focus on the technical issues of energy efficiency.

Our use of energy in modern life includes powering appliances and vehicles and in cooling and warming structures. As energy costs have become more important and visible, we have already seen impressive progress in designing more energy-efficient air conditioners, refrigerators, coffee makers, computers, and other electronic devices. To some degree, it has simply been a matter of adding energy efficiency as a design specification. While the work required of the electric device provides a technical parameter for the amount of energy needed to perform a function, this too has been subject to technical innovation. Compare the light-emitting diode to the incandescent bulb. According to the wonderful Web site *HowStuffWorks.com*: "LEDs are just tiny light bulbs that fit easily into an electrical circuit. But unlike ordinary incandescent bulbs, they don't have a filament that will burn out, and they don't get especially hot. They are illuminated solely by the movement of electrons in a semiconductor material, and they last just as long as a standard transistor" (Harris 2002). The site also notes that the main advantage of LEDs is that they use less energy than other forms of illumination:

> In conventional incandescent bulbs, the light-production process involves generating a lot of heat (the filament must be warmed). This is completely wasted energy, unless you're using the lamp as a heater, because a huge portion of the available electricity isn't going toward producing visible light. LEDs generate very little heat, relatively speaking. A much higher percentage of the electrical power is going directly to generating light, which cuts down on the electricity demands considerably.
>
> (Harris 2002)

Similar innovations have made many household appliances more energy efficient. Building structures, insulation, ventilation, heating, cooling, water supply, and sewage treatment systems have all been made more energy efficient and show promise for becoming even more efficient in the future. Technologies utilizing sensors and timers are able to turn devices on and off to save energy. While there are examples of home technologies that are surprisingly energy *in*efficient, such as cable TV boxes, we find that as people learn that they are wasting energy needlessly, they look for ways to save energy if possible. Heating and cooling loss from buildings can now be assessed by low-cost thermographic scans and other energy audit techniques (Fiedler et al. 2009).

Finally, there is the issue of energy-efficient transportation. Of course, the most energy-efficient form of transportation is walking or cycling—very low-tech solutions. Other technologies include mass transit and auto engines such as hybrids and electric vehicles. There is little question that personal and mass transit can improve their energy efficiency. In the first decade of the twenty-first century, we began to see efforts to develop these technologies.

Because we waste so much energy in the United States, there are many examples of energy-saving practices that are relatively easy and inexpensive to implement. This low-hanging fruit requires little technical innovation and provides enormous practical potential for sustainability when coupled with the new technologies that are being developed.

Financing Renewable Energy and Energy Efficiency

Just as America has transformed itself from an industrial economy to a higher-end information- and service-oriented economy, we are now at the start of a transformation to a green economy. In June 2009, the Pew Foundation released a study on this transformation, which concluded:

> The number of jobs in America's emerging clean energy economy grew nearly two and a half times faster than overall jobs between 1998 and 2007. . . . Pew developed a clear, data-driven definition of

> the clean energy economy and conducted the first-ever hard count across all 50 states of the actual jobs, companies and venture capital investments that supply the growing market demand for environmentally friendly products and services.
>
> (McGillis 2009)

Climate and energy policy can help modernize our economy's technological base and ultimately increase our standard of living. Our goal should be to ensure that the percentage of our Gross Domestic Product (GDP) devoted to energy expenditure is as low as possible. According to the United States Energy Information Agency of the Department of Energy, there has been a fair amount of volatility in this indicator over the past forty years. The first year that the federal government reported our expenditures on energy as a percent of GDP was in 1970. That year, we spent 8 percent of the GDP on energy. This grew to 11.6 percent in 1979 and peaked at 13.7 percent in 1981. In 1999, it dropped to an all-time low of 6 percent, because of a fast-growing economy and low fuel prices. However, in the twenty-first century, this percentage has tended to grow. It jumped to 7 percent in 2000 and to 8.8 percent in 2006, the last year for which we have government data (EIA 2009).

In 2009, with the economy shrinking and fuel prices increasing, the amount of the nation's wealth devoted to energy grew once again. The question we need to address in the long run is: how do we reduce the price and also the unpredictability of energy costs? Fossil fuels are subject to a wide variety of unpredictable cost factors—ranging from the increased use of automobiles in China to the vagaries of Middle East politics. In the long run, however, the cost of fossil fuels is bound to grow. While the Earth retains huge quantities of fossil fuels, we are not making any more of the stuff. Each day that we burn fossil fuels, less remain. Fossil fuels will continue to get more difficult and expensive to extract, and the environmental impact of extraction will not let up. While fuel extraction can be made more cost effective and environmentally friendly through the use of technology, the fundamentals remain: fossil fuels will become more expensive over the next century.

In contrast, look at the cost of computing. According to Moore's Law, a truism first popularized by Gordon Moore, one of the founders of Intel, the processing power of microchips doubles every eighteen months, and the cost of computing drops every year. Anyone who buys a laptop knows that they keep getting more powerful and less expensive. Solar technology has the potential for the same type of cost reductions over time. The basic fuel of solar power, the sun, will always cost the same to tap—nothing. Solar cells and batteries will only get less expensive as a mass market develops and technology improves.

The long-term costs of renewable energy and energy efficiency are lower than those of fossil fuels and wasteful energy practices. However, some forms of energy efficiency and all forms of renewable energy require capital and involve higher costs in the short and middle term. These fundamental issues of sustainability finance were introduced in chapter 2 and are certainly applicable to energy. In the case of energy efficiency, public policy has tied electric utility self-interest together with the public policy objective of energy efficiency. Electric utilities are highly regulated private businesses. In exchange for a local monopoly, they accept high levels of regulation. In some states, these utilities no longer see their profits tied to increased levels of electrical consumption and are actually rewarded for serving more customers with less power (Pew Center on Global Climate Change 2009). Utilities seeking to avoid the capital and political costs of building new power plants have begun to focus more attention on energy efficiency. As mentioned earlier, a surcharge on electricity imposed by the states of California and New York has generated substantial sums of money that can be used to subsidize energy efficiency practices and equipment in businesses and households. Government programs such as these can influence the cost-benefit calculations of energy users and encourage them to take steps to become more energy efficient.

Financing renewable energy is a more complex problem. The first problem is that fossil fuel prices are relatively low and do not include the price of the external damage they generate. As we have learned in recent years, the price of fuels such as gasoline tends to be quite volatile. A tax on carbon or serious regulation of global warming gases will have the effect of raising the prices of fossil fuels and will make renewable energy more cost competitive. Of course, if the price of energy grows too quickly, it can have a negative effect on the economy.

By reducing the cost of energy, financing capital investments in renewable energy and energy efficiency will eventually be part of the normal costs of doing business. In his typically well-thought-through piece "More Profit with Less Carbon," the noted environmentalist Amory Lovins (2005, 74) observed:

> Over the past decade, chemical manufacturer DuPont has boosted production nearly 30 percent but cut energy use 7 percent and greenhouse gas emissions 72 percent (measured in terms of their carbon dioxide equivalent), saving more than $2 billion so far. Five other major firms—IBM, British Telecom, Alcan, NorskeCanada and Bayer—have collectively saved at least another $2 billion since the early 1990s by reducing their carbon emissions more than 60 percent. . . . These sharp-penciled firms, and dozens like them, know that energy efficiency improves the bottom line and yields even more valuable side benefits: higher quality and reliability in energy-efficient factories, 6 to 16 percent higher labor productivity in efficient offices, and 40 percent higher sales in stores skillfully designed to be illuminated primarily by daylight. . . . The U.S. now uses 47 percent less energy per dollar of economic output than it did 30 years ago, lowering costs by $1 billion a day.

Many companies have become more receptive to energy efficiency initiatives and are incorporating them as a mainstay of their carbon reduction strategies. Jill Jusko (2009, 1), a senior editor at *Industryweek*, writes that

> 71% of respondents to a recent survey commissioned by the International Facility Management Association and Johnson Controls reported that they are paying more attention to energy efficiency than they were one year ago. And of organizations that have made public carbon commitments, 45% have identified energy efficiency in buildings as their top carbon-reduction strategy.

Jusko observes that funding is the main barrier to energy efficiency projects. More upfront capital is required for renewable energy than for conventional energy projects. In the book *Feed-in Tariffs: Accelerating the Deployment of Renewable Energy*, Miguel Mendonca (2007,

3–4), a research manager for the World Future Council, observes that renewable energy projects are characterized by

> high initial capital costs. Even though lower fuel and operating costs may make renewable energy cost-competitive on a life cycle basis, higher initial capital costs can mean that RE provides less generation capacity per initial dollar invested than conventional energy sources. Thus, RE investments generally require higher amounts of financing for the same capacity. Depending on the circumstances, capital markets may demand a premium in lending rates for financing RE projects because more capital is being risked up front than in conventional energy projects. RE technologies may also face high taxes and import duties. These duties may exacerbate the high first-cost considerations relative to other technologies and fuels.

Mendonca also writes that renewable energy projects are typically smaller and, because of new technology, more complex than conventional projects. They therefore have higher transaction costs per unit of energy generated. He cites this as a barrier to the growth of renewable energy. He observes that

> the transaction costs of RE projects—including resource assessment, siting, permitting, planning, developing project proposals, assembling financing packages and negotiating power-purchase contracts with utilities—may be much larger on a per-kW capacity basis than for conventional power plants. Higher transaction costs are not necessarily an economic distortion in the same way as some other barriers, but simply make renewables more expensive. However, in practice some transaction costs may be unnecessarily high—for example, overly burdensome utility interconnection requirements and high utility fees for engineering reviews and inspection.
>
> (Mendonca 2007, 5)

The renewable energy industry is still not fully understood and tested, and so investor confidence has been slow to develop. Renewable energy projects are often viewed as being high risk because of their financial structure and scale. The absence of accurate and standardized information about renewable energy projects is a barrier to growth:

> Renewable energies, although subject to the same market forces as conventional energy sources, involve markedly different technologies and thus their financing requires new thinking, new risk management approaches and new forms of capital. Half the battle for renewables is to instill confidence within the investment community, which will happen only after financiers have traveled up the learning curve that gives them exposure and understanding of the real risks and opportunities associated with financing RE.
>
> (Sonntag-O'Brien 2004)

While renewable energy faces capital constraints that are easy to understand, energy efficiency also faces capital limits, despite its relatively short payback time. In 2009, the United Nations Environment Program issued a report entitled "Global Trends in Sustainable Energy Investment 2009." The report's executive summary concluded that venture capital and private investment was growing in the green energy sector:

> New investment in biofuels reached $16.9 billion, down 9% from 2007. Other renewable energies such as geothermal and minihydro were up 26% to $5.4 billion, but there was a 25% fall in investment in biomass and waste-to-energy to $7.9 billion. Private investment in new energy efficiency technologies was $1.8 billion, a fall of 33% on the previous year. However, the energy efficiency sector recorded the second highest levels of venture capital and private equity investment (after solar), which will help companies develop the next generation of sustainable energy technologies.
>
> (UNEP 2009, 12)

An interesting development in electric utility regulation is that rate setting has been used to encourage energy efficiency. In the past, utility profits and therefore rate of return on investment increased along with the use of electricity. Utilities are now rewarded for investments in energy efficiency infrastructure and with higher rates when they demonstrate that their customers have lowered utilization because of efficiency gains.

There are barriers to implementing inexpensive energy efficiency technologies. Among them are the behavior of landlords, builders, and consumers. Typically, landlords and builders have a short-term mindset and invest in low-cost equipment that does not include energy-efficient components, even though in the long run more efficient equipment would pay off. Often the problem is that tenants and homeowners pay the costs of energy, while landlords and builders pay the cost of equipment.

As the cost of energy grows and as organizations strive for greater efficiency and lower levels of waste, energy costs are being seen as potential areas of saving. Moreover, since the technology used is generally off the shelf, the return on investment is predictable as long as the price of energy is relatively stable. In this respect, energy efficiency is simpler to finance than renewable energy. While the argument can be made that we don't need to develop renewables because we have lots of fossil fuels left, no one is in favor of wasting energy. Financing efficiency is bound to seem less risky than financing new energy technologies.

The United Nations Environment Program's report on trends in sustainable energy investment observed that while the wind and solar sectors have traditionally been characterized by supply chain bottlenecks, increases in scale and operating experience have made clean technology less costly. The UNEP (2009, 12) report notes:

> During 2008, wind was the largest sector in terms of new investment, while solar took second place by surpassing biofuels. Total financial investment in wind was $51.8 billion, down 1% on 2007, and in solar was $33.5 billion, up 49% from the previous year. A large proportion of this investment went into wind and solar projects, particularly in the established markets of the European Union and North America, but also increasingly in China, Eastern Europe and Latin America. Sustainable energy technologies on the whole are becoming cheaper to manufacture as they reach scale and gain operating experience. Recently, this has not always translated into price decreases because of demand outstripping supply and commodity prices soaring. But the investment surge of recent years and softened commodity markets have started to ease supply chain bottlenecks, especially in the wind and solar sectors, which will cause prices to fall towards marginal costs and several players to consoli-

date (at the end of 2008 there were over 70 major wind turbine manufacturers globally and over 450 photovoltaic (PV) module makers). The price of solar PV modules, for example, is predicted to fall by over 43% in 2009.

It is clear that economics can be both an incentive and a deterrent for making the switch to sustainable energy sources. In the following case studies, again based on blog posts I wrote for the *Huffington Post*, I further discuss the possibilities of financing the sustainable energy sector. My discussions focus on the Property Assessed Clean Energy (PACE) program. The first, "Financing Green Energy," addresses this potential public policy solution and its ability to help homeowners save money while reducing their environmental impacts. The second, "The Federal Government Attacks Creative Local Green Energy Finance," discusses of the federal government's unfortunate attack on this positive and effective means of financing green energy.

CASE STUDY: FINANCING GREEN ENERGY

In order to develop a green and sustainable economy, we need efficient and non–fossil fuel based energy production and consumption systems. A number of technological innovations are needed before we can implement a smart grid and proliferate the use of solar energy. But a great deal can be done now if we take advantage of existing technology.

We are all aware of the need to make the transition from a fossil fuel–based economy to a renewable and more energy-efficient one. While we will not run out of fossil fuels in my lifetime or (probably) yours, finite resources will only get scarcer, more expensive, and harder to get out of the ground. Even if you reject the compelling science of climate change, what real argument is there against a cheap, secure energy supply and efficient energy use? That the old ways are better?

We need to develop and implement a way to make green energy profitable. This is easier said than done. First of all, we need a lot of money. According to a United Nations 2009 report on sustainable energy finance, global investments will have to reach five hundred billion dollars annually for the next ten years if we want greenhouse gas emissions to stabilize and start declining after 2020.

Five hundred billion dollars is no small amount, and the private sector will have to play a more prominent role than governments in the transition to a sustainable energy economy. Thanks to the slow return rate on green investments, not to mention the recession, progress has been slow.

The main difficulty with financing green energy in the private sector is that investments require lots of cash, and returns are uncertain. Landlords, for example, have little incentive to invest in energy-efficient retrofits that will accrue savings to their tenants but not to themselves. This problem is particularly pronounced where rents are strictly regulated, like in New York City. Similarly, homeowners may not see a return on their investments before they intend to sell a property and likewise have no incentive to invest in products such as insulation or double-glazed windows. Investors will save money over the long run—but these days, who cares about the long run?

Renewable energy suffers from the same affliction. Even though the costs of renewable energy will go down as the technology develops, financial returns are uncertain, and given the choice, private equity firms are more likely to fund a fossil fuel project than a renewable energy project. Many investors think that the technology is still too new and risky.

However, change is coming, and where the private sector falls short, public policy can help. One example of an innovative method of financing green energy is Property Assessed Clean Energy, or PACE legislation. PACE was one of the topics discussed in a Columbia University panel discussion that I moderated recently (Earth Institute 2010). PACE provides communities with the ability to designate a district where property owners can finance energy efficiency or renewable energy improvements to their property. The debt is then added to the property tax bill until it is retired, so the costs of the improvements stay with the property if the current owner sells. As of today, sixteen states, New York included, have passed PACE-enabling legislation.

Until the financial recession, there were strong upward trends in private investment in green energy production. In 2008, investment in renewable energy was, for the first time, greater than new investment in fossil fuels. That year, the total transaction value for the renewable energy sector—both public and private—was over 220 billion dollars. That's not five hundred billion, but it is an indication of the growth of the green energy field.

Growth in the renewable energy sector fell as the financial markets destabilized during the financial crisis. In the first quarter of 2009, there was less than half the amount of private investment in green energy projects than during the same period of 2008. However, while the private credit market fell victim to the recession, the world's governments stepped up. In 2008, governments around the world, including the government of

(*continued*)

CASE STUDY: FINANCING GREEN ENERGY (*continued*)

the United States, allocated a total of 155 billion dollars to green energy programs. Even as the recession deepened, governments passed stimulus packages increasing funding for green energy to 180 billion dollars, signaling the strong political support for efficiency and renewable energy investments.

Both the public and private spheres have shown their support for green energy. Despite this progress, investments in green energy are still considered risky by many financial professionals, and recessions make all of us more risk averse. It is now obvious that we need innovative ways to help steer funding toward energy efficiency and renewable energy production projects, despite fluctuations in the availability of credit. PACE is one innovative method. Others will surely follow.

CASE STUDY: THE FEDERAL GOVERNMENT ATTACKS CREATIVE LOCAL GREEN ENERGY FINANCE

The financial geniuses at Fannie Mae and Freddie Mac who managed to lose billions of dollars in the housing market have decided in their newfound fiscal conservatism to do their best to derail a promising and innovative mechanism for financing local green energy retrofits. The federal housing agencies' attack in July 2010 on an energy- and money-saving program, Property Assessed Clean Energy (PACE), demonstrates an obtuse hostility toward green energy initiatives.

PACE programs allow local governments to sell municipal bonds and lend the capital to local homeowners for energy efficiency and renewable energy projects. This solves two problems that usually stop homeowners from retrofitting older homes for green energy: the availability of capital and the possibility that the time it takes to recoup the energy-cost savings from your investment could be longer than the amount of time that you own your home. PACE programs add the green energy loan repayment to a homeowner's property tax bill, which means that the costs of energy-saving investments are assumed by new homeowners if the house is sold. After ten to twenty years, the additional assessment ends once the loan is repaid. The PACE program provides a great incentive for homeowners to switch to renewable energy or reduce their energy consumption.

Unfortunately, federal financial institutions object to the PACE program. In an article written this past March, the *Wall Street Journal*'s Nick Timiraos (2010) outlined the financial dilemma posed by this new mechanism when he noted that PACE "debt would be senior to existing mortgage debt, so if the homeowner defaults or goes into foreclosure, it would be repaid before the mortgage lender gets any money. While property-tax assessments are usually senior to existing property debt, cities have traditionally used their assessment authority for community-wide improvements like sewers and roads—not for upgrades that homeowners elect to make on their own homes. Proponents of the program, called Property Assessed Clean Energy, or PACE, say it is necessary for the loans to be paid before mortgages if local governments are to raise funds for the program from municipal-bond investors."

At the heart of these financial institutions' objections to PACE is a shortsighted and narrow view of the world that willfully ignores the value of reducing the cost of energy in the home. Rather than working with state and local governments to ensure that PACE investments add value to the home in ways that address potential financial objections, these financial wizards simply say they will not buy or sell mortgages that include PACE-related liens. In a *New York Times* article on June 30, 2010, Todd Woody reported: "In letters sent to mortgage lenders on May 5, Fannie Mae and Freddie Mac stated that energy-efficiency liens could not take priority over a mortgage. 'The purpose of this industry letter is to remind seller/servicers that an energy-related lien may not be senior to any mortgage delivered to Freddie Mac,' wrote Patricia J. McClung, a Freddie Mac executive. However, the agencies did not offer guidance to mortgage lenders on how to handle properties that carry the energy liens. Backers of the programs fear that mortgage lenders, who depend on Fannie and Freddie to buy their home loans, will now start demanding that the entire lien be paid off before issuing a new loan."

It is true that many mortgages are in trouble because homeowners have taken out too much debt on their homes and have borrowed and spent their housing equity on frivolous purchases like big-screen TVs and hot tubs. However, a more efficient furnace, insulation, and solar panels are investments that increase in value as the price of energy rises. PACE programs can be designed to require certified energy audits, higher levels of financial capability, and other restrictions. These lumbering financial giants should not be allowed to destroy this promising local initiative.

But nobody is stopping them. In a statement issued on July 6, 2010, the Federal Housing Finance Agency exempted homes with existing PACE loans from proposed restrictions but directed Fannie and Freddie to:

> Undertake actions that protect their safe and sound operations. These include, but are not limited to:

(*continued*)

CASE STUDY: THE FEDERAL GOVERNMENT ATTACKS CREATIVE LOCAL GREEN ENERGY FINANCE (*continued*)

- Adjusting loan-to-value ratios to reflect the maximum permissible PACE loan amount available to borrowers in PACE jurisdictions;
- Ensuring that loan covenants require approval/consent for any PACE loan;
- Tightening borrower debt-to-income ratios to account for additional obligations associated with possible future PACE loans.

(FHFA 2010)

The Federal Housing Finance Agency also encourages Fannie and Freddie to develop additional guidelines for these green energy finance programs. While this new approach is far better than automatic rejection, it may be a way of replacing the initial sledgehammer elimination of PACE with death by a thousand small cuts.

What is most disturbing about the approach taken by these federal housing agencies is their obvious hostility to the goals of green energy. While they provide lip service in support of green energy goals, their actions speak louder than their words. As of today, twenty-two states have authorized PACE programs, but these are fledgling efforts at best. At this crucial early stage and in the very complex world of home finance, housing agencies have done what I am afraid may be permanent damage to a promising initiative. If PACE is to be saved, the Obama administration needs to send a clear message to all federal agencies, including those bailed-out, quasi-private bodies like Fannie and Freddie, to get on the green energy train. Today.

Developing the Organizational, Institutional, and Technical Capacity to Meet the Management Challenges of Renewable Energy and Energy Efficiency

Even when funds can be found to finance energy efficiency or renewable energy, it is not always easy to identify organizations with the capacity to implement these energy projects. Even while capacity is growing worldwide, it is dwarfed by the capacity devoted to fossil fuels. In the case of energy efficiency, a number of companies are growing to manage, deliver, and implement energy efficiency services.

On the implementation side, a large number of energy service companies, or ESCOs, are gradually taking root across the United States. According to the National Association of Energy Service Companies' Web site:

> An ESCO, or Energy Service Company, is a business that develops, installs, and arranges financing for projects designed to improve the energy efficiency and maintenance costs for facilities over a seven to twenty year time period. ESCOs generally act as project developers for a wide range of tasks and assume the technical and performance risk associated with the project. Typically, they offer the following services:
>
> Develop, design, and arrange financing for energy efficiency projects;
> Install and maintain the energy efficient equipment involved;
> Measure, monitor, and verify the project's energy savings; and
> Assume the risk that the project will save the amount of energy guaranteed.
>
> These services are bundled into the project's cost and are repaid through the dollar savings generated. . . . What sets ESCOs apart from other firms that offer energy efficiency, like consulting firms and equipment contractors, is the concept of performance-based contracting. When an ESCO undertakes a project, the company's compensation, and often the project's financing, are directly linked to the amount of energy that is actually saved.
>
> (NAESCO.org 2009, 3)

A May 2007 study by the Lawrence Berkeley National Laboratory and the National Association of Energy Service companies indicated that ESCO industry revenues grew by 20 percent from 2004 to 2006, to a total of $3.6 billion in 2006 (Hopper et al. 2007, v).

This model has many advantages for small businesses that do not have sufficient capital, but it encountered some problems during the financial crisis of 2008 and 2009, when some of the ESCOs capital sources dried up, making them unable to finance large-scale efficiency projects. A second problem with this model is that ESCOs will work to ensure that energy efficiency projects generate short-term dollar savings, because that is how they make money. This can constrain the type of projects that are proposed. Often, they are best used as part of a more comprehensive energy efficiency strategy, one either generated internally or by a firm that focuses on developing but not implementing energy efficiency strategies.

Over the past several years, I have worked as a senior advisor to such a firm, the California-based Willdan Energy Solutions. This company is not an ESCO but is typically funded by electric utilities to work with energy consumers, to develop strategies and management practices that can improve energy efficiency. In some of their projects, they help organizations identify ESCOs and work with them.

While the organizational capacity for energy efficiency is challenging, the capacity to develop renewable energy is even more difficult. Here the problem relates to the difficulties often faced by high-tech startup firms. Capital is often inadequate, standard operating procedures are still under development, and a focus on creative invention can reduce the attention devoted to the more prosaic tasks related to implementation.

In some cases, the structure of the American system of energy generation and distribution creates problems for new renewable energy generators. One issue is insurance. There are many insurance gaps for covering renewable energy risk, and many instruments do not address the unique needs of the industry. This is a barrier to growth. A key problem for insurers is assessing risk for new and untested technologies. A second issue is that renewable energy sources may not be compensated for the full value of the power that they supply to the electrical grid. Mendonca (2007, 4) observes that:

> RE sources feeding into an electric power grid may not receive full credit for the value of their power. Two factors are at work here. Firstly, RE generated on distribution networks near final consumers rather than at centralized generation facilities may not require transmission and distribution (i.e., would displace power coming from a transmission line into a node of a distribution network). But utilities may pay only wholesale rates for the power, as if the generation was located far from final consumers and required transmission and distribution. Thus, the "locational" value of the power is not captured by the producer. Secondly, RE is often an "intermittent" source whose output level depends on the resource

(i.e., wind and sun) and cannot yet be entirely controlled. Utilities cannot count on the power at any given time and may lower prices for it.

Mendonca further notes that there may be unique building restrictions on renewable energy projects: "Wind turbines, rooftop solar hot water heaters, photovoltaic (PV) installations and biomass combustion facilities may all encounter building restrictions based upon height, aesthetics, noise or safety, particularly in urban areas" (Mendonca 2007, 5).

A lack of transmission lines connecting energy sources to urban areas is a major technical capacity issue. Capacity may be restricted because of transmission access and utility interconnection requirements. The current system of power generation does not promote renewable energy technologies but rather fossil-fueled technologies. However, new generation capacity derived from renewable energy is increasing. According to the UN Environment Program:

> Combined with approximately 25GW of new large hydropower stations, renewable energy overall represented 41% of total new global capacity. 2008 was the first year that investment in new power generation capacity sourced from renewable energy technologies (approximately $140 billion including large hydro) was more than the investment in fossil-fueled technologies (approximately $110 billion). Given the long life of power sector assets, however, it will be some time before renewable energy dominates the generation mix. In 2008, renewable energy still only accounted for 6.2 % of total power sector capacity.
>
> (UNEP 2009, 11)

Another key capacity issue is the availability of skilled labor. Many of the technologies of both energy efficiency and renewable energy are new, and they require workers trained to understand and operate these new technologies. Many of the companies involved in the renewable energy business are small, undercapitalized, entrepreneurial startups. While some major corporations have dipped a toe tentatively into the renewable energy business, with the exception of a few major players such as the wind farm advocate T. Boone Pickens, the efforts in the United States have been modest. That has resulted in equally modest capacity building. Interestingly enough, the world economic crisis of 2008–2009 had the effect of stimulating the renewable energy business in China, which had the effect of increasing organizational capacity in both China and the United States. In *The New York Times* in August 2009, Keith Bradsher (2009) wrote that:

> China . . . has stepped on the gas in an effort to become the dominant player in green energy—especially in solar power. . . . Backed by lavish government support, the Chinese are preparing to build plants to assemble their products in the United States to bypass protectionist legislation. . . . Since March [2009], Chinese governments at the national, provincial and even local level have been competing with one another to offer solar companies ever more generous subsidies, including free land, and cash for research and development. State-owned banks are flooding the industry with loans at considerably lower interest rates than available in Europe or the United States.

At the close of the first decade of the twenty-first century, the organizational capacity of the renewable energy industry was still in its infancy. The price of solar power was typically not competitive with fossil fuels. Seeing this, but also understanding the long-term promise of solar, the central command-and-control government of China targeted solar as an export industry and invested state resources in subsidies for this new industry. Of course, most governments are unable to operate in the single-minded, large-scale way that the Chinese government does. The concluding section of this chapter focuses on the role of government and public policy in stimulating energy efficiency and renewable energy.

The Role of Government and Public Policy

There is often controversy and difficulty in identifying and implementing a role for government in the development of a new industry. For some analysts, such as J. P. Painuly of the UNEP Collaborating Centre on Energy and Environment, there is no issue. Painuly believes that the government can promote renewable energy through a variety of initiatives:

> Imperfections and distortions in the market coupled with unfavourable financial, institutional and regulatory environments imply that governmental intervention is not only desirable but also a must to promote RETs [renewable energy technologies]. . . . The role includes generic actions to remove barriers, building human and institutional capacity, setting up research and development infrastructure, creating an enabling environment for investment, and providing information and mechanisms to promote RETs.
>
> (Painuly 2001, 84)

He also argues for energy sector liberalization in an effort to open the energy sector to more intense competition:

> Some examples of the specific policies are: creating separate entities for generation and distribution in the electricity sector, allowing private sector entry and diluting or removing controls on energy pricing, fuel use, fuel import, and capacity expansion etc. . . . The basic purpose of liberalisation is to increase efficiency of the energy sector through facilitating market competition.
>
> (Painuly 2001, 85)

Bringing competition to certain parts of the energy business is complicated by the public utility tradition often found in some of these industries. In the case of the distribution of gas and electricity, it has long seemed sensible to avoid the duplication of gas pipelines and power lines. This resulted in the development of highly regulated quasi-monopolies for both energy generation and distribution. In recent years, power generation has been opened up to competition in some places by requiring distribution lines to accept electricity from more than one source. However, one of the obstacles to decentralized renewable energy is the varied set of rules governing access to the power grid.

While facilitating competition is one role for government policy, both energy efficiency and renewable energy can be brought about more rapidly if government is willing to play other roles as well. In the case of energy efficiency, a degree of government policy is already in place, and more is on the way. Policies have been set to require buildings, vehicles, and appliances to meet energy efficiency standards. Programs funded by surcharges on utility bills have successfully encouraged best management practices. In states such as California and

New York, these funds provide small businesses with funding for new equipment and retrofits to reduce wasteful energy practices (CPUC 2010). In the case of renewable energy, tax credits and other policy instruments are used to encourage the development and use of renewable energy facilities.

About 80 percent of New York City's greenhouse gases come from the city's buildings. On Earth Day 2009, New York's mayor Michael Bloomberg and City Council Speaker Christine Quinn announced the "Greener Greater Buildings Plan for NYC." The city set the goal of cutting 30 percent of its carbon emissions by 2030. According to the press release, the city would require buildings of fifty thousand square feet or more to conduct an energy audit once every ten years and make any improvements that pay for themselves within five years (Navarro 2009).

Unfortunately, the city council scaled back the plan before it was enacted. The audits were still required, but not the efficiency upgrades. The problem was that the real estate industry could not figure out any way to recoup their investments. Building owners would pay the costs of improved efficiency, but under some leases they would not be able to recover their costs, and only tenants would receive the benefits. And, even if some owners were willing to invest and could recoup, private capital became increasingly scarce in 2009, and the resources were simply not available (Navarro 2009).

While New York City was scaling back its initiative, many people in Washington, D.C., were pushing a "Cash for Caulkers" program, in part as a jobs initiative. Under this program, federal stimulus money would be paid to homeowners to fund 50 percent or more of the costs of a weatherization project (Hargreaves 2009). The greater use of existing technology is the low-hanging fruit of energy sustainability, and energy efficiency is the least controversial practice in the field of energy policy. It is hard to argue against energy efficiency—what would you say: "I am in favor of waste and spending money that I could save"? However, government policy is needed to facilitate investment in technology and practices that enhance efficiency.

While the greater use of existing technology is a sound short-run strategy, in the long run new technology must be developed and then commercialized. Government must fund the basic research and development of new renewable energy and energy efficiency technologies. It must also provide incentives for investment in the twenty-first-century "smart grid" electricity transmission system. A number of analysts and elected leaders, including President Barack Obama, have called for a "moon-shot" project to develop renewable energy technology. This is a focused effort, funded by government, to develop a few specific technologies. As you may have guessed, my favorite idea is that we improve the technology and cost effectiveness of solar cells and solar batteries.

Solar power is in its infancy. In 2007, the U.S. Department of Energy selected thirteen cities to help build the country's solar energy market. As part of this partnership, New York City set a goal of increasing its photovoltaic cell capacity from 1.1 MW in 2005 to 8.1 MW by 2015. This is, of course, a small drop in a very large bucket. According to ConEd.com (2010):

> New York's energy use has reached unprecedented levels. For the year 2007, Consolidated Edison Company of New York's customers used 62,591 gigawatt hours (GWh) of electricity, which eclipsed the previous record of 61,608 GWh set in 2005. This level of use is more than 23% higher than the 50,837 GWh used in 1997. A gigawatt is a rate of energy production equal to 1,000 megawatts. According to the latest available national data, Con Edison's record delivery surpasses the annual electrical usage of the entire state of Colorado (49,734 GWh in 2006) or the Commonwealth of Massachusetts (55,850 GWh in 2006).

I provide these data only to communicate a sense of scale. One gigawatt equals one thousand megawatts. The tiny contribution of solar power is dwarfed by the growth of electrical demand in New York City. Still, you've got to start somewhere.

There are two ways to measure electrical power use: annually (as we did earlier) and based on peak demand. Peak demand is important, because you have to provide enough power to meet demand when everyone wants to use it, just as a shopping mall's parking lot needs enough spaces for the day before and after Christmas. In the power business, you need enough power for the hottest day in August. Both peak load and annual use are growing in New York City.

Why is residential power use growing? It is partially because of population growth but largely because of the growth of the number of electrical devices in our homes. The increasing number of computers, air conditioners, iPods, CD players, TVs, microwaves, and video game systems are increasing our need for power.

So why is solar making such a pathetic contribution to meeting our power needs? Is solar power for real? The short answer is yes. Solar cells are coming down in price, and government tax incentives, higher oil costs, and fears of global warming are all contributing to the growth of solar power.

New York City provides some challenges to the use of solar power that other cities do not present. It is in some ways a worst-case scenario for solar. While most of the land in New York City sits beneath single-family homes, most of the people in New York City live in apartment buildings (Department of City Planning, City of New York 2010; Starr 1988). Many of our apartment buildings do not have the space and sunlight needed for current solar technologies to provide enough power to warrant investment.

However, the technological base for solar power is starting to change. We saw this happen to computing over the past fifty years. While we once needed an entire room to house a computer that had less memory than a typical laptop, a revolution in miniaturization has shrunk the world of electronic devices. Many engineers think this will soon happened to photovoltaic technology. Despite the nearly complete absence of federal funding for solar energy research, some of our best scientists and engineers are working to improve solar cells.

In fact, in February 2008, G. Pascal Zachary reported in the *New York Times* that a number of Silicon Valley's chip designers are now working on solar cell technology. That piece noted that both solar and computer chip technologies were silicon based and that to some chip engineers, solar cells were really a type of "chip." Some solar enthusiasts see solar power as inevitable and cite its impressive recent growth rate as evidence that we will soon be living in a solar-powered world. Even without new technology, other analysts see room for an increased use of solar energy in our energy mix (Cohen 2008b).

I agree that solar energy has enormous potential. My engineering colleagues at Columbia tell me that the Earth absorbs much more energy in the form of sunlight than we could ever need to power our homes and businesses. The problem is that we don't know how to efficiently collect that energy and store it. The technology of solar cells must become more efficient and practical, and the power we take from the sun must be stored in a more cost-effective battery.

Imagine if all of the power used in your home could be fueled by a set of solar cells that could fit on a single windowpane. If you think it's impossible, imagine it's the year 1950, and someone tells you that some day you will carry a telephone in your pocket that is smaller than a wallet—and that it would work anywhere. Or that you will carry five thousand songs and 150 movies in a machine that is no larger than that tiny phone. You would think that these ideas were far-fetched if not a little crazy.

The technological transformation of miniaturization and satellite communication began with the needs of national defense and the space program. The same was true of the Internet, which began as a Department of Defense effort to share research data across facilities. The commercial development of these technologies by the private sector followed rapidly but would have never taken place without the initial massive government investment in the basic technology. Such investment will be needed to ensure energy sustainability going forward. The following "memos" explore how potential energy savings can be measured through energy audits.

CASE STUDY: ENERGY EFFICIENCY, ENERGY AUDITS

What They Are

To begin an exploration of energy efficiency options, it is important to first discover where the most opportunities for savings exist. An energy audit is a method to determine the amount of electricity being used by each appliance in a building and how much of it is being used productively. For example, an energy audit could reveal that a television is using an excess of electricity when on standby or that heat is escaping a home from under the doors. By pinpointing electricity use by appliance and points of inefficiency, it becomes much simpler to find low-hanging fruit for reducing energy use (U.S. DOE 2009).

How They Work

There are several ways to undertake an energy audit. Numerous downloadable kits are available through government energy agencies' Web sites, and some firms specialize in performing detailed energy audits of businesses and homes.

(continued)

CASE STUDY: ENERGY EFFICIENCY, ENERGY AUDITS (*continued*)

Undertaking a basic energy audit can be surprisingly simple. A do-it-yourself audit focuses mainly on climate control, an important energy use in nearly all buildings. The audit usually involves detecting drafts and poorly insulated places in the home by feeling around outlets and the frames of doors and windows. A professional energy auditor may use equipment such as blowers and infrared cameras to more accurately detect places where warm (or cool) air is escaping the building. Many utility companies and government agencies offer free or low-cost energy audits (U.S. DOE 2009). New York's Con Edison is beginning to provide free energy audits to their mid-sized commercial customers and to a limited number of residences. I expect this service to increase in the next several years.

As they have increased in popularity, the equipment to undertake energy audits has become readily available. Electric usage monitors, small devices that connect an appliance and an outlet and display in real time how much energy is being used, are now for sale at many electronics stores. By estimating the amount of time that the appliance is needlessly in use and multiplying it by the electricity usage per second, one can find the energy wasted by that appliance per time period. With this information, one can then determine if changes in use patterns would create substantial energy savings.

Energy use dashboards are an increasingly popular form of energy audit. These employ electric usage monitors throughout the home that broadcast a wireless signal to a central monitor, which displays electricity use graphically on a computer. This allows the homeowner or building operator to observe and control how much electricity any appliance is using at any time and has been shown to yield average energy reductions of 15 percent through changes in user behavior (Lohr 2008; see also the "Smart Grid Technology" section, below).

Who Benefits?

Government: Any entity that operates large numbers of buildings can benefit from energy audits, and governments are no exception. In New Jersey, the State Board of Public Utilities has created an incentive program to pay 75 to 100 percent of the costs of a contracted energy audit for schools, hospitals, courtrooms, fire stations, and other publicly owned buildings. This program has created a way for the government to save tax dollars and put into practice the energy efficiency measures it encourages in private industry and citizens (New Jersey Clean Energy Program 2009).

Businesses: Energy audits can often result in substantial financial savings, which is particularly important for small businesses. At the family-operated Goody, Goody restaurant in St. Louis, Missouri, an energy audit led to savings of over $1,300 per year since 1998. The restaurant took advantage of an offer by their utility company to perform an energy audit of the establishment. The resulting report pinpointed potential savings by changing lighting fixtures to more efficient models, installing motion sensors in storage areas, and other low-cost energy-saving strategies. The total cost of implementing the report's recommendations was only six hundred dollars—an amount earned back in about half a year (U.S. EPA 2009).

Homeowners: There are numerous examples of successful energy audits for residences. One such audit took place in Salisbury, Massachusetts. A homeowner found that their newly built house would not retain warmth in the winter or coolness in the summer. An energy audit undertaken through the Energy Star program at the U.S. Environmental Protection Agency using fans and infrared cameras located the problem not just with the wall insulation, as the homeowner had suspected, but with leaky and uninsulated ducts in the basement. As a result of the energy audit's report, the homeowner was able to apply for grants to help pay for the retrofits needed to reduce their home's energy use. After the grants, the cost to the homeowner was only five hundred dollars. The retrofits saved over two hundred dollars per year, meaning that the homeowner's return on investment was only 2.5 years.

Challenges

Despite the many benefits of energy audits, they are not without risk to the consumer. While utilities audits are nearly always accurate and useful, private companies sometimes undertake fraudulent energy audits. These often involve unnecessary procedures or false findings intended to lead to unnecessary purchases (Kaupp 2008).

It is difficult to regulate energy audits, since the consumer is the only agent who can truly determine whether the audit was successful, and then only after a time lag. However, a certification process for legitimate energy auditors might be one answer. Because fraudulent auditors may be more financially successful in the short run, there is actually a danger that they will outcompete honest auditors, leading to poor quality throughout the marketplace. Regulatory challenges remain in the area of energy audits (Kaupp 2008). In addition, even if waste is detected, the funds to make a home or business more energy efficient may not be available.

CASE STUDY: THE ENERGY STAR PROGRAM

Any initiative for reducing our dependence on fossil fuels must include investment not only in alternative forms of energy but also in reducing overall energy use. Initiatives such as the Energy Star program, operated by the U.S. Environmental Protection Agency and the U.S. Department of Energy, facilitate purchases and practices that reduce energy use through labeling. This strategy allows customers to take an active role in reducing their carbon footprints and is a simple, market-oriented way to reduce energy use.

What the Program Does

The Energy Star program is a system for evaluating and labeling appliances, practices, and even buildings to meet energy efficiency standards set by the U.S. government. The label informs customers that a certain appliance or building is more energy efficient than other available options (U.S. EPA 2010b).

Energy Star began in 1992 as a labeling program for computers, and it has since expanded to include thirty-five types of goods. The logo has become a recognizable symbol for millions of customers. According to EPA surveys, 75 percent of Americans recognize and understand the Energy Star symbol. In addition to labeling, the program provides free, downloadable toolkits for energy management and compiles lists of consulting companies that can provide direct assistance to businesses and homeowners who want to reduce energy use. Overall, the EPA estimated that the program saved energy customers $19 billion in 2008 alone (U.S. EPA 2010b).

How It Works

To evaluate whether they should develop standards for a category of products that use energy, the EPA uses several criteria, including the probability of nationwide savings. This means that any product that shows significant energy efficiency relative to its competitors per unit may be considered, even if sales are small. Similarly, products that have large sales but only see low relative efficiency per unit will also be considered (U.S. EPA 2010b).

Other criteria include maintenance of product quality, potential for a customer to recover any additional cost of the good, and the ability for multiple technologies or approaches to achieve the standard. This last is particularly important, since it contributes to the market neutrality of the program. No company is unduly favored over another; all have the opportunity to meet the specifications (U.S. EPA 2010b).

To set the standards for a particular category of goods, the Energy Star technicians monitor the products and reevaluate as needed. The process includes the extensive participation of environmental groups as well as partnership with the companies that manufacture and market the products (Consumer Reports 2008).

Who Has Benefited

Because of the versatility of the initiative, many different stakeholders have had the opportunity to participate in and benefit from the Energy Star program. Some success stories from various sectors are discussed below.

State government: Computers and monitors are major users of energy in a modern office, and they are often sold without their energy efficiency functions enabled. The State of Delaware Department of Natural Resources, with the help of the Delaware Department of Energy and the EPA, initiated a program to encourage all their offices to enable the "sleep" functions for their computers and monitors. This action alone can save up to fifty dollars per computer, totaling over $14,000 annually for the Delaware Department of Natural Resources. The initiative inspired many employees of the state office to enable these functions in their home computers as well (U.S. EPA 2010a).

To create additional incentives for computer energy savings, the Energy Star program includes the Million Monitor Drive, a program that encourages private and public partners to collectively enable power-saving functions on one million monitors per year. The result is substantial energy and financial savings as well as positive publicity for participating organizations (U.S. EPA 2010a).

City government: The city of Portland, Oregon, has long been upheld as a model for sustainable municipal management, and part of their strategy has been to pursue energy efficiency both to prevent emissions and to save money for the city. In 2001, Portland's municipal government, faced with increasing electricity rates, took advantage of utility companies' rebates and low technology costs to replace over six thousand incandescent traffic lights with LED-based signals (Energy Division, City of Portland 2001). LEDs have a threefold longer lifespan than incandescent bulbs (six years, as opposed to two) and use only about one-seventh of the energy. The city of Portland recovered their costs within five years and continues to see

(*continued*)

CASE STUDY: THE ENERGY STAR PROGRAM (*continued*)

savings of nearly $400,000 per year relative to their previous maintenance costs (Energy Division, City of Portland 2001). While the Energy Star program was not directly involved in this project, they promote similar initiatives in other cities and have rated and labeled LED traffic lights to encourage their use (Energy Division, City of Portland 2001).

Businesses: Many vendors of Energy Star appliances have found that participation in the program can increase their sales and attract new customers. As new regulations come into place and electricity rates increase, customers begin to actively seek energy-efficient appliances, and Energy Star labels provide an easy way to identify them.

For one lighting fixture vendor in California, two strategies proved particularly successful. First, she created showrooms devoted specifically to Energy Star fixtures, which attracted customers and increased her sales dramatically. She also hosted an "Energy Star breakfast" with local homebuilders and other potential customers to educate them about her inventory of energy-efficient lighting apparatus. The response was overwhelmingly positive, and her sales again grew. Her experience is an excellent example of how effective a reliable labeling mechanism can be for vendors as well as customers (U.S. EPA 2010a).

Universities: Like the Delaware Department of Natural Resources, the University of Wisconsin found that they were using excessive resources to power their computers. However, the need for nightly maintenance created an additional factor; they could not put the computers into "sleep" mode without disabling their ability to accept new software updates (U.S. EPA 2010a).

However, with the help of free downloadable software provided through the Energy Star program, they were able to allow their computers to enter sleep mode, wake automatically for updates, and then go to sleep again afterward. With a free program and a few hours of staff time, they saved an average of twenty dollars per computer, totaling $9,000 per year (U.S. EPA 2010a).

Challenges

The program is not without flaws. Like many such labeling schemes, it has faced criticism from environmental and consumer advocates for not being sufficiently stringent. Some groups, such as Consumer Reports, found that certain refrigerators that were Energy Star rated consumed twice as much electricity as indicated by the EPA.

Another objection to the program is that the standards do not always relate to normal usage patterns. The fundamental flaw, Consumer Reports claims, is the lack of an independent audit of the Energy Star testing procedures. For example, they may test a refrigerator for energy consumption while leaving the icemaker off, which is unlikely to occur in a normal household. The Department of Energy also often relies on industry reporting of competitors' products as opposed to conducting their own tests (ConsumerReports.com 2008).

The Natural Resource Defense Council, an environmental advocacy group, has criticized the program for allowing too many appliances to bear the label. They stated that 92 percent of dishwashers in 2006 were Energy Star labeled, while ideally only the top 25 percent should earn the label (Aston 2008). Despite these flaws, the creation of a trusted and recognized label is certainly a step toward reducing the nation's carbon footprint.

CASE STUDY: SMART GRID TECHNOLOGY

Revamping America's energy distribution system is a crucial component of our shift toward a cleaner energy economy. New and promising technologies for distributed generation, including combined heat and power, solar panels, windmills, and energy efficiency necessitate a flexible system of energy transfer. Startup firms, established utilities, and consumers are all contributing to the shift toward a new energy economy, and the federal government has played a significant role in facilitating this evolution (U.S. DOE 2009).

Why the Smart Grid Is Needed

The infrastructure that transports energy from producers to customers, referred to as the grid, worked fairly efficiently when energy flows were one way, from producer to customer, and when energy supply far exceeded demand. However, the emerging complexities of the energy market have revealed the weaknesses of the grid's one-way flow of information and electricity. Smart grid projects, which the U.S. Department of Energy is mandated to support and undertake as part of the Energy Independence and Security Act of 2007, increase the ability of customers to interact with producers through infrastructural improvements to the physical grid (U.S. DOE 2009). The result will be an energy delivery system with low transaction costs that delivers power where and when it is needed, leading to more efficient outcomes.

(*continued*)

CASE STUDY: SMART GRID TECHNOLOGY (*continued*)

How It Works

A smart grid is a system that links sensors, placed throughout the physical system of wires and substations, to the Internet or a similar network. This allows information concerning energy load and need throughout the grid to be monitored and for both utilities and customers to respond to the information in ways that improve energy efficiency. For example, a utility provider can increase electricity rates at peak hours, accounting for their equipment strain during that time. A consumer, seeing the price change on their personal energy dashboard at their computer terminal, can respond by remotely changing the temperature on their thermostat at home, reducing their own energy use (Lohr 2008).

Other benefits include facilitating distributed generation. A utility in a sunny state can know how much to change the electricity load provided during the hottest part of the day, when air conditioners are operating, to account for increased energy production by solar panels on rooftops. A smart grid would also include capacitors for storing energy for use during peak hours, a key component to integrating renewable sources such as wind and solar, which do not produce a constant supply (U.S. DOE 2009).

By reducing strain on equipment and increasing the options for energy production, smart grid technologies have the potential to decrease the need for environmentally costly energy production facilities. In addition, investment in this new infrastructure has the potential to create hundreds of thousands of jobs and prevent costly power supply disturbances (U.S. DOE 2009).

Current Efforts and Studies

The Pacific Northwest National Laboratory, operated by the U.S. Department of Energy, undertook one of the first consumer tests of smart grid technology in 2007. Electric use controls and sensors were installed in 112 homes in Washington State and connected to the Internet. Each household had a Web portal through which they could set their energy use preferences concerning the range of temperatures they would accept for their home climates, along with their acceptance level for different electricity rates. On the portal, rates and electricity use was presented visually, not numerically, to facilitate user understanding (Lohr 2008).

Throughout the study, the customers actively engaged with their energy use, both directly and through automatic changes made according to their preset preferences. These changes in energy use created minute-to-minute changes in rates, similar to a stock exchange. The prices reflected the real time usage of energy and stimulated changes in consumer behavior in response. One customer lamented the loss of the equipment after the study ended, indicating the level of personal enjoyment he found in his ability to participate in the energy market (Lohr 2008).

As a result of their use of the smart grid technology, the customers saved 10 to 15 percent from their energy bills per year. If this equipment could be provided throughout the country, savings of that magnitude could prevent billions of dollars of spending on new energy production facilities and help prevent air pollution, climate change, and other environmental damage (Lohr 2008).

A small industry is hard at work commercializing these technologies. Companies that work on smart grid technologies generally sell the equipment to utilities, which then distribute it to their customers and have access to the customers' use information. Only some of these apparatus display the use data to the consumer as well as the utility. However, Google has recently introduced equipment that displays the use data to both the customer and the producer, similar to the type tested by the Department of Energy (LaMonica 2009).

Companies engaged with smart grid technology take their cues from the telecommunications industry. For instance, cellular phone companies create incentives to reduce peak usage by offering free minutes during nights and weekends. This reduces the load on their infrastructure and saves the customers' money on calls that could be made at any time. Other possibilities include allowing prepaid plans for low-income electricity customers with bad credit, just as cell phone companies offer. Prepay systems for electricity are already in use in student housing in Europe (Lento 2009).

Challenges

While the smart grid idea has gained great popularity throughout the United States, there are still several hurdles to overcome. With the advent of monitoring software, the problem of compatibility emerges. If utility companies sell to their customers equipment that is only compatible with their monitoring software, then the customer cannot switch to another utility without having to purchase new equipment. Google has advocated for the development of regulations to standardize the monitoring, so that each utility does not create a metering format that is incompatible with others (LaMonica 2009).

The large size of the investment in expanding and upgrading our electrical grid presents an obstacle as well. While many private firms have

(*continued*)

CASE STUDY: SMART GRID TECHNOLOGY (*continued*)

emerged to capitalize on the transition, government research, grants, investment, and incentives have and continue to play an important role in expediting the evolution of smart grid technologies (U.S. DOE 2009).

Despite these obstacles, the combined initiative of the public and private sectors have improved and proliferated smart grid technology only in the last decade. As the technology becomes widely adopted throughout the country, customers and producers alike will reap the benefits of positive environmental and economic outcomes.

Chapter 4

Sustainable Water

Definition and Technical Challenges

Although water is a finite resource on our planet, when humans and animals use it, it doesn't disappear. It either evaporates into the atmosphere and returns in the form of rain or it combines with other chemicals, making it less fit for human consumption. Most of the water on the planet is in the form of salt water, which doesn't quench our thirst but turns out to be pretty helpful for many species of fish. The focus of this chapter is on maintaining the quality and quantity of water needed for human life and the support of key ecosystems and food supply systems.

There is some disagreement among scholars about the presence of a global water crisis, but despite that disagreement there is little question that with the human population of the planet continuing to grow, many traditional sources of water have been damaged. My Columbia colleagues Upmanu Lall, Tanya Heikkila, Casey Brown, and Tobias Siegfried have divided the water issue into three components: (1) the issue of access to water because of the absence of infrastructure to store, treat, and deliver it; (2) the problem of damaged or polluted water; and (3) the issue of water scarcity resulting from imbalances of supply and demand in some geographical settings (Lall et al. 2008).

These issues of access, pollution, and the geographic imbalance of human population and water supply seem to me to present a succinct definition of the component parts of the challenges to freshwater sustainability.

Water Supply: Quantity, Distribution, and Efficiency

With the planet approaching seven billion people and heading toward a probable population of ten billion, the question of water *quantity* arises: is there a sustainable amount of potable water that can withstand the increasing demand because of climate change and population growth? Worldwide, about 70 percent of the freshwater we use goes to farming, 20 percent is used by industry, and only 10 percent is used in households.

According to Frank Rijsberman (2006, 1), in his article "Water Scarcity: Fact or Fiction?": "Water will be a major constraint for agriculture in coming decades and particularly in Asia and Africa this will require major institutional adjustments." Increased demand for food will be the main driver of freshwater scarcity in the next several decades. While household water is also likely to be scarce in some parts of the world, the water demands of agriculture are known to be high, and therefore it is water for that purpose that presents the greatest challenge to sustainability.

There is disagreement among experts about the absolute shortage of water. The hydrologic system is closed—no water vapor leaves the planet, and most of the planet is covered by water. It is difficult, however, to measure the amount of potable water present on the planet. Freshwater can be combined with other substances, making it less useful for people or their crops. The United Nations projects that two-thirds of the world's population will experience water stress by 2025 (UNEP 2002).

With a majority of the people of the planet now living in cities and a growing number of people in the developing world moving into urban areas, it is the more rapidly developing cities that will more likely experience water scarcity. In part, this is because of inadequate infrastructure and the capital required to build waterworks, and in part this is because of the unequal distribution of water resources throughout the world (UNESCO 2003).

As the developing world rapidly urbanizes, inadequate capital for infrastructure is not limited to water. Transportation, waste, sewage, energy, education, and health facilities are also scarce, and many people live in appalling slums and shantytowns. According to Ellen J. Lee and Kellogg J. Schwab (2009, 109):

> Rapidly growing populations and migration to urban areas in developing countries has resulted in a vital need for the establishment of centralized water systems to disseminate potable water to residents. Protected source water and modern, well-maintained drinking water treatment plants can provide water adequate for human consumption. However, ageing, stressed or poorly maintained distribution systems can cause the quality of piped drinking water to deteriorate below acceptable levels and pose serious health risks. . . . Distribution system deficiencies in developing countries [are] caused by: the failure to disinfect water or maintain a proper disinfection residual; low pipeline water pressure; intermittent service; excessive network leakages; corrosion of parts; inadequate sewage disposal; and inequitable pricing and usage of water.

Generating the capital for a centralized water system is a challenge in the developed world, and it is even more difficult in the developing world. Systems to treat and distribute water and collect and treat sewage are essential to public health and welfare and will somehow need to be constructed in every city of the developing world. In the developed world, these facilities must be upgraded and maintained as well. Over the past several decades, residents and businesses in New York City have directly paid the costs of this infrastructure in the form of rapidly rising water bills. The result of that spending is detailed in the case study below, originally published on the *New York Observer*'s Web site in 2008.

CASE STUDY: THE GOOD NEWS ABOUT NEW YORK CITY'S WATER

With the furor over the economy and the financial crisis, it's easy to forget that we are actually capable of acting like a real community and building for the future. I say "sometimes," because, while this city has a magnificent system for delivering fresh water to its people, it has one of the worst solid waste management systems imaginable. Let's focus on the good news: New York City's water supply system.

(*continued*)

CASE STUDY: THE GOOD NEWS ABOUT NEW YORK CITY'S WATER (*continued*)

New York gets its water from two upstate reservoir systems that it owns and operates. To keep the sources of water clean, the city works upstate to purchase land and ensure best management practices by local farmers and other residents. According to the New York City Department of Environmental Protection's 2006 water supply report, "the Department of Environmental Protection (DEP) has developed a $19.5 billion Capital Investment Strategy for the next decade, the majority of which will be used to upgrade and add to existing infrastructure and guarantee that we can fulfill our mandate of delivering quality drinking water to New York for years to come" (Department of Environmental Protection, City of New York 2006, 1).

New York's water system provides more than 1.1 billion gallons of water daily to around eight million New York City residents and one million residents in Westchester, Putman, Ulster, and Orange counties (Saucier 2008). The two tunnels that carry our water to us represent one of the most impressive public works projects in the world. Water Tunnel no. 1 was completed in 1917, Water Tunnel no. 2 was completed in 1936, and construction on Water Tunnel no. 3 began in 1970 and with luck will be completed in 2020. According to a water industry Web site: "New York's City Tunnel No. 3 is one of the most complex and intricate engineering projects in the world. Constructed by the New York City Department of Environmental Protection, the tunnel will eventually span 60 miles and is expected to be complete by 2020" (Water-technology.net 2010).

One reason we are building a new water tunnel is in the hope that over the next century we can repair the other two tunnels. Some experts estimate that about a third of the water that we draw from our upstate system leaks before it gets to our faucets. In fact, since the late 1980s, the Delaware Aqueduct, a piece of vital infrastructure that carries half of the city's water, has been leaking between 10 and 36 million tons of water each day (Saucier 2008). The city is not waiting for the third water tunnel to be completed to plug this leak—a new project was just started to fix this problem (Saucier 2008).

While we may lose a lot of our supply, the quality of our water is quite good. As Elizabeth Royte wrote in her wonderful 2007 *New York Times* piece, "On the Water Front": "The upstate water is of such good quality, in fact, that the city is not even required to filter it, a distinction shared with only four other major American cities: Boston, San Francisco, Seattle, and Portland, Ore. New Yorkers drink their water from Esopus Creek, from Schoharie Creek, from the Neversink River, straight from the city's many reservoirs, with only a rough screening and, for most of the year, just a shot of chlorine and chasers of fluoride, orthophosphate and sodium hydroxide" (Royte 2007, 1).

The city's filtration exemption from the EPA saves it from the cost of building a six-to-eight-billion-dollar water filtration plant for the water that comes from the Catskill and Delaware watersheds located west of the Hudson River. It would cost about one billion dollars a year to pay the debt service and operating costs of that plant (DePalma 2007). A majority of our water comes from west of the Hudson. The rest of our water comes from the Croton Watershed up in Westchester and Putnam counties. The city is currently spending two billion dollars to build a water filtration plant under the Mosholu Golf Course in the Bronx to protect our water supplies that come from east of the Hudson (Office of the Comptroller, City of New York 2009).

The city is working hard to protect the waters that it doesn't need to filter. According to the former commissioner of New York's Department of Environmental Protection, Emily Lloyd, "In order to preserve this remarkable asset, and prevent the need for an expensive filtration plant for the Catskill and Delaware water systems, the city enforces an array of environmental regulations designed to protect water quality while encouraging reasonable and responsible development in the watershed communities. It also invests in infrastructure—such as wastewater treatment facilities and septic systems—that shield the water supply, while working with its upstate partners to develop comprehensive land-use practices that curb pollution at the water's source" (Department of Environmental Protection, City of New York 2010).

The city has spent over one billion dollars during the past decade in the communities near the water supply to keep development from ruining the water. This turns out to be cheaper than the billion dollars per year that a filtration plant would cost (DePalma 2007).

Most of New York City's water supply is protected and filtered by the natural processes of upstate ecosystems. To environmental economists, nature's work that protects our water is an "environmental service." Because the price of a filtration plant is known, we can estimate the monetary value of the services provided to filter our water. This comes to one billion dollars per year, minus the hundred million or so we spend each year to protect the upstate ecosystems. This is nine hundred million dollars a year of found money that we will lose if we don't protect our fragile ecosystems. It's a graphic illustration of the point that what is good for the environment will often be good for our bank account. Sustainable development is more than a slogan—it is a principle of good government and sound fiscal management. New York's water is a good news story that will only stay good if we pay attention and protect it from harm.

The Technology of Water

With enough capital and energy, the world's supply of fresh water could be expanded dramatically. Some of that expansion is well underway. In many parts of the United States, golf courses are already reusing treated wastewater to irrigate fairways and greens. Each year, more and more of our water is filtered before we use it, and more and more sewage is treated as well.

However, the capital requirements for modern sewage and water treatment facilities can run into the billions of dollars for construction and tens of millions per year in operation and maintenance. The EPA estimates that typical construction of a wastewater facility for a community of fifty thousand costs $12.5 million, and plants serving larger communities can be far more costly. The Blue Plains facility in the Chesapeake Bay watershed, for example, has received a court order to improve its facility to prevent sewage overflows, at an estimated cost of $2 billion (U.S. EPA 2008). A single sewage processing plant in New York City has annual operating costs of $100 million (Department of Environmental Protection, City of New York 2007).

Many areas that once would have been too arid for human settlement have benefited from the development of massive waterworks projects. Without the multistate, multi-billion-dollar Colorado River project, the modern cities of Phoenix, Las Vegas, and Los Angeles could not have been possible, as they all depend on aqueduct systems that draw from this source of water. The possibility of long-distance transport and long-term storage permits human settlements to take advantage of water supplies that are intermittent and located far from where people decide to live.

The filtration of water for reuse and desalinization are both technologies that could vastly increase the supply of potable water. However, these technologies are still being perfected and are both capital and energy intensive. In this sense, if low-cost renewable energy could be developed, these water treatment technologies would come down in price and become more cost effective as well (Zimmerman et al. 2008).

Efficient Use of Water

One way to sustain the level of water needed is to increase overall water productivity (Rijsberman 2006). This means wasting less water in households, but more critically, it means learning to use less water to grow our crops. In nations with modern industrial agriculture, such as the United States, over 70 percent of water utilization is devoted to agriculture (Zimmerman et al. 2008). Agricultural water use practices are often wasteful, with only 30 to 40 percent of the water used in open canal irrigation reaching the target plant (Associated Press 2006). In part, this inefficient behavior can be traced to government management. High-cost government subsidy programs are very common in both developed and developing countries (La Vina 2006). Many of these programs provide incentives for farmers to grow only a few specific crops and can have the effect of discouraging farmers from choosing crops based on local conditions, including water availability. The result is intense pressure on water supplies in arid and semiarid areas to grow water-intensive crops such as wheat and rice. Other programs actively subsidize irrigation water, artificially depressing prices and resulting in poor water management (La Vina 2006).

While there are many known technological methods for reducing water use in agriculture, the entire system of what is grown where would need to be rethought to achieve higher levels of water efficiency (Associated Press 2006). Among other steps, crops would need to be grown where there is sufficient rain for their water needs. Irrigation methods could also be made more efficient. The EPA recommends the following methods for reducing agricultural water use: "chiseling of extremely compacted soils, furrow diking to prevent runoff, and leveling of the land to distribute water more evenly." The EPA also notes that "typically, field practices are not very costly" (U.S. EPA 2010c).

Household use of water can be reduced by installing low-flow toilets, faucets, and showerheads. Efforts at public education can also help reduce wasteful water use. Of course, if the technology of water reuse becomes less expensive and more common, water efficiency will become less important, as more water will be recycled.

Water Quality

When thinking about water sustainability, we need to think about water pollution and water quality as well. There are many definitions of water quality, and different uses require different levels of quality and purity. The main factors used to determine water quality include the water's clarity, salinity, total suspended solids, chemical contaminants, biological contaminants such as fecal matter, and dissolved oxygen. Different uses of water can tolerate different levels of water quality. Water is used for drinking, bathing, agriculture, recreation, and industry (U.S. EPA 2007). Poor water quality is not a trivial issue; it is a major source of health problems in the developing world. Nearly one billion people do not have access to drinkable water, and over 2.5 billion people do not have access to improved sanitation (World Health Organization 2009). Infectious diseases can result from water polluted by human and animal waste, and cancer can result from exposure to toxic chemicals dumped in waterways after industrial use. In rural areas, water is polluted by runoff from farms that have been treated with herbicides, pesticides, and fertilizer (Gadgill 1998).

While much of this chapter focuses on the issue of sustainable freshwater, it is worth mentioning the issue of ocean pollution. The ocean has long been used as a place to dump human waste streams, and as those waste streams have become less biodegradable, we have seen the accumulation of increasing amounts of marine debris. Everything from the plastic that holds six-packs of beer to shopping bags has been added to the daily load of "floatables" that wash up on to our beaches and are ingested by wildlife every day (Derraik 2002).

Additionally, a loss of biodiversity has resulted from overfishing and the practice of inadvertently introducing invasive species of life via the ballast and wastewater of ships. Water taken in from a freshwater source on one continent may be dumped with waste into a waterway on another continent. These water sources include life forms that are alien to the ecosystems they end up in and can damage these other systems (U.S. EPA 2010d). This type of damage is not limited to the oceans but also affects larger freshwater supplies as well.

The American Great Lakes, for example, have been overwhelmed with invasive species. In fact, according to a study team of masters students at Columbia University:

> The United States Geological Survey has identified 136 species of invasive algae, fish, invertebrates, and plants that have established themselves in the Great Lakes since the 1800s. These species have caused substantial ecologic and economic damage to the region. Specific attention is given to the Zebra Mussel (*Dreissena polymorpha*), a freshwater mollusk native to the Caspian Sea, responsible for up to $500 million per year in economic and environmental damages.
> (Srivastava et al. 2009, 6)

Another key issue related to water quality is the condition of the facilities and pipes that carry the water from its source to the place where it is finally used. Pumping stations and pipes could be contaminated or in a state of disrepair, and filtration systems could be poorly operated or maintained. In some cases, the water pipes in old apartment buildings may leach contaminants into the water supply (Center for Disease Control 2010).

Water quality is, of course, related to water quantity; having a great deal of contaminated water is less useful than the possession of a smaller quantity of high-quality water. Both water quality and water quantity can be enhanced with good management and sufficient capital to purchase appropriate technology.

Organizational, Management, and Financial Challenges

Water collection, transportation, filtration, and storage are among the most capital intensive of human activities (Neukrug 2001). Sewage and wastewater collection and treatment is equally complex and expensive. Both water supply and sewage treatment are essential to human well-being. While many of the basic technologies have been in place for many years, the past several decades have brought substantial improvement in the technologies of water supply and sewage treatment. This means that organizational outputs tend to be better than they once were. However, each new technology is accompanied by the requirement for new standard operating procedures, new training, and inevitable mistakes.

Because of technological advances and increased regulatory requirements, many places have a mix of older and newer technologies.

This means that staff must be trained to work in a variety of facilities. This increases the challenges to management as they struggle to understand and successfully manage the operations of a water facility. Some of the earlier sewage treatment facilities in the United States produced a sludge that was simply dumped at sea. Under new rules, the dumping of waste is now illegal, and fortunately plants can now provide advanced (tertiary) treatment to sewage. To meet the new requirements, the sludge from old plants is simply barged to the newer plants and treated there. Transporting sludge by barge is both time consuming and expensive—but presumably less expensive than building a new plant or adding more advanced treatment to a current facility (Department of Environmental Protection, City of New York 2010).

The organizational capacity of running a water or sewage treatment facility should never be assumed or taken for granted. Old technology requires familiar and experienced hands to be run effectively, and new technology often requires that staff learn new ways of operating equipment. Mistakes can be costly and even threatening to human health. We rely on a series of complex urban systems to deliver power, water, and transportation. Equally complex systems are in place to remove solid waste, storm water, and sewage. We tend to take these systems for granted and assume that they are self-implementing. In fact, they require a great deal of planning, coordination, and capital to be developed, improved, and maintained.

Financing sustainable water and sewage systems can be a challenge for the wealthiest communities. As the case study on New York City's water system discussed, water supply systems are expensive. New water tunnels and water filtration facilities are multi-billion-dollar investments. In order to maintain New York City's water supply, the state of New York created a Water Finance Authority and Water Board in 1984. According to New York City's official Web site:

> The New York City Municipal Water Finance Authority . . . provides funding through the issuance of bonds, commercial paper and other obligations to finance the capital projects required to ensure the continued supply and purity of the City's high quality drinking water and for safe wastewater collection, treatment and disposal. . . . The Authority, together with the New York City Water Board and

the New York City Department of Environmental Protection, manage the City's water and wastewater system. The Department of Environmental Protection operates and maintains the system with a work force of over 5,600. The Water Board, which was created by the New York State legislature at the time of the creation of the Water Authority, has the primary responsibility to levy and collect water and wastewater rates and charges.

(Municipal Water Finance Authority 2010)

Over the past several decades, water supply has become a more important issue in New York City and in cities all over the world. One result of this was that the cost of water has grown for businesses and households. In many places, the household water bill has become an established fact of life in the United States. While in earlier, simpler times, water may have come from local wells and was free of charge, today that has become increasingly rare. Just as TV was once a free broadcast service, for many the cable bill, water bill, and cell phone bill are now routine and expected. And just as cable and cell phone prices are increasing, water bills too are growing. In 1985, the water rate in New York City was one dollar per hundred cubic feet. By 2010, it was nearly seven dollars—a rate of increase far in excess of inflation (Water Board, City of New York 2010).

The price of water grows as additional technology and energy are used to purify and transport it. New processes are required to treat sewage and filter water. These can be energy intensive and expensive. As the planet gets more crowded, the infrastructure of water transport must be placed in more densely settled areas. This process makes land acquisition and construction more complicated and expensive. In the developed world, the public's concern for its health and overall quality of life creates increased pressure on governments to deliver environmental amenities such as clean water. People and businesses are more mobile than ever. If one place is unable to provide clean water, people will move to a place that can deliver this resource. Therefore access to low-cost capital is necessary for the construction and reconstruction of water treatment facilities. A dedicated revenue stream is also required in order to pay the debt service on the capital and to fund the operation and maintenance of the system. In an increasing number of places, user fees and other forms of dedicated taxes provide the necessary revenue.

All of this brings us back to the issue of organizational capacity: what are the specific capacities required to ensure sustainable water? The types of skills required to deliver water and treat sewage include hydrology, civil engineering, finance, project management, and community relations, to name a few. An understanding of the area's ecology and water resources is essential, as is a detailed historic memory of the water infrastructure. Each system will have some common features but will tend to have important attributes unique to its geographical, social, economic, and political environment. The skills required are not commonplace, and system failures or breakdowns are more frequent than we would like to admit.

In some places, localities have given up on their effort to develop and maintain this technical capacity. They contract with the private sector for elements of their water and sewage system, or in some instances they may seek a contractor to run the entire system. This privatization of water facilities has succeeded in some places, such as Indianapolis, and failed in other places, such as Atlanta. Interestingly, in those two cases, the same company, Suez Lyonnaise des Eaux/United Water, was the contractor (Cohen and Eimicke 2008, 197). The ideological argument that either government or the private sector is better at running water utilities is worth avoiding. Neither sector has a monopoly on competence. In some cases, government can manage the system with heavy contractor support from private vendors. In other cases, government agencies may find it easier and cheaper to do the whole thing in-house. In some places, such as Indianapolis, privatized systems are extremely successful. In the majority of places, we see a mix of government and private organizations working under the direction of public decision makers. The specific mix that is most effective will vary by location and should be the subject of ongoing strategic analysis.

Conclusions

There is no question that water supply is a critical responsibility of government. Our need for water is not debatable. The skills of the private sector are an essential part of the sustainable water equation. Generating capital, managing infrastructure construction and operation, and practical technological innovation are all best performed by the private sector. On the other hand, basic research and development, quality regulation, and the regulation of those monopolies that by necessity must emerge in treating and distributing water are all governmental functions.

Just as energy can be used far more efficiently, water can also be used more efficiently. Given the large proportion of the water supply devoted to agricultural use, it is obvious that we should focus our attention on reducing water usage in agriculture. Crops can be assessed for their use of water, and losses due to the transport of irrigation water can be reduced. The technology of waste treatment, water filtration, and desalinization can also be used to effectively increase our supply of fresh water. These technologies tend to be very energy intensive, and thus a successful search for lower cost and renewable energy sources would also improve our ability to meet our water needs.

Chapter 5

Sustainable Food Supply

Definition and Technical Challenges

Most U.S. cities have daily or weekly farmers' markets. These are places in a park or on a street where local farmers come one or two days per week to sell local fresh produce, baked goods, meat, and dairy products. I asked a city official who worked with the neighborhood farmers' market if the very popular program was going to expand further over the next few years. His response was that they could not, because the area was "out of farms," meaning that just about all of the farms within 150 miles of the city were already operating booths in the markets. If we take New York City as an example, it is clear that a city of over eight million in a metropolitan area of over twenty million cannot be fed by local sources of food. This is not to argue that community gardening and local food do not have an important role to play in our food system. It simply means that most of the job will require a sustainable agriculture industry.

With the world's population becoming more urban every year, the distance between people and their sources of food will continue to grow. This means that what has been termed "industrial farming" will become more important and that we need to focus greater attention on the sustainability and environmental impact of our farming practices.

According to the Union of Concerned Scientists, "industrial agriculture views the farm much as a factory, with inputs and outputs, with the goal of increasing yield and decreasing cost of production" (UCS 2007). As a result, small farms have consolidated into larger units, and traditional methods of what we might call "family farming" have been replaced by factory-style production aimed at satisfying consumer demand for low-priced, plentiful food.

A by-product of industrial agriculture has been increased impacts on local ecology and the global biosphere. Industrial farming results in increased energy use for food growth, refrigeration, and transport, as well as increased use of fertilizer and pesticides in farming. Runoff from these agricultural factories damages local groundwater and surface water supplies in some places and causes habitat and species destruction in others (Matson et al. 2007). Pollution control and reduced energy use are challenges that the agriculture industry is slowly beginning to address.

Our dependence on industrial agriculture for food production cannot be understated, and so there is an urgent need to make the agricultural industry more sustainable. In the second half of the twentieth century, the industrial system of food production and distribution became dominant. According to the World Food and Agriculture Organization (FAO), between 1950 and the late 1990s, the amount of cultivated land increased by 18.7 percent, and the world average yield per hectare rose 90 percent (FAO 1999). According to the Pew Commission on Industrial Farm Animal Production, "since 1960, milk production has doubled, meat production has tripled, and egg production has increased fourfold" (Pew Charitable Trusts 2008, 5).

While industrial agriculture has been incredibly successful in increasing food production, it has resulted not only in environmental damage but human damage as well. The issue of hunger in the developing world is complex but incredibly real and painful. The uneven and unpredictable distribution of "industrial" food in the developed world coupled with damage to local soils in the developing world results in unpredictable local farming and a literally a situation of feast and famine (Edwards 1989).

While most developing countries have seen a decline in the number of malnourished children, that problem is increasing in sub-Saharan Africa. The number of hungry people in sub-Saharan Africa increased 20 percent from 1990 to 2000. According to the United Nations Capacity Building Task Force on Trade, Environment, and Development: "In the period 2000–2002, the proportion of undernourished people in the total population of Kenya was 33 per cent, in Uganda 19 per cent and in the United Republic of Tanzania 44 per cent. . . . The majority of the chronically hungry are small farmers in developing countries" (UNEP 2008, vii).

The planet is capable of growing sufficient food for the approximately seven billion people that now inhabit it. It could even grow the food supply needed to feed ten billion people. The current problems are linked to distribution, wealth, and the unsustainable agricultural practices now in place. These are not intractable problems. The issue of ecological management is critical. As indicated over two decades ago by the agricultural scientist C. A. Edwards (1989, 27), "the use of agrochemicals can increase yields dramatically for a period but without good management there is rapid soil deterioration and erosion when land is cropped for extended periods."

There are a large number of destructive agricultural practices that must be modified through the implementation of sustainable agriculture policies because of the environmental problems they create (Crosson 1986). These destructive practices include the use of pesticides, the overuse of fertilizer and the use of chemical fertilizers, poorly managed irrigation infrastructure, the conversion of forests to pasture or crops, and overcultivation or nutrient mining.

Pesticide Use

One of the early chemical pesticides banned by law in the United States was DDT. Its threat to human health became well known, but its effectiveness against malaria-carrying mosquitoes made its banning controversial and resulted in frequent violations.

Pesticides have resulted in fish kills, human poisonings, the loss of wildlife, and pest adaptation to the chemicals (Crosson 1986). Nevertheless, the use of pesticides has improved crop yields in many cases, and the pressure to find safer but effective pest-reduction agents continues (Eccleston 2008).

Fertilizer

When fertilizer is applied to a farm's field crops, water carries off whatever is not absorbed by the plants being fertilized. These chemicals, most of which are nitrogenous, can have a major and negative effect on freshwater supplies—especially groundwater (Sing 1979). By overloading the nutrients in the water, eutrophication and dead zones result from oxygen-depleted water (Pant et al., 1980). Another effect is that over time the soil loses its ability to contain plant nutrients and becomes less fertile (Rasool et al. 2007). Advances in the technology of fertilizer content and application hold out the promise for reduced environmental impact, but in many cases, state-of-the art-techniques are either unknown, ignored, or beyond the capacity of the farmer (Kahlown et al. 2006).

Irrigation

Irrigation can solve the problem of water supply in more arid regions of the world but can create problems as well. A frequent problem results from saltwater intrusion when irrigation projects are located near oceans or other bodies of saltwater, or, most often, when the groundwater itself sits on a saline layer. When this occurs, the heavier saltwater rises to the surface as the lighter freshwater is pumped away (Smets et al. 1997). Once freshwater is contaminated by saltwater, it becomes very difficult to filter for human consumption. Another problem, especially in tropical climates, is that the incidence of water-associated diseases increases in proximity to these artificial waterways (Samadpour et al. 2005).

Conversion of Forests to Pastures and Croplands / Nutrient Mining

One of the most significant results of cutting down trees is that when the land has been cleared for grazing or crops, the soil begins to erode. Many ecologists consider the American Dust Bowl of the 1920s and 1930s as an ecological disaster stimulated by overfarming the land. When we replace complex and interconnected ecosystems with monocultures, we create an environment more vulnerable to attack. On the other hand, industrial agriculture, like all agriculture, requires land. It is important that the soil being used for food production is periodically measured and analyzed to ensure that it retains the capacity to grow food. Overcultivation of land can result in its degradation and can reduce its productivity. This is sometimes called "nutrient mining." According to Julio Henano and Carols Baanante (2009, 2):

> During the 2002–2004 cropping season, about 85% of African farmland (185 million hectares) had nutrient mining rates of more than 30 kg/ha of nutrients yearly, and 40% had rates greater than 60 kg/ha yearly. . . . About 95 million hectares [about 7.7 times as big as New York State] of soil have reached such a state of degradation that only huge investments could make them productive again.

As I indicated earlier, our population size and urban lifestyles require industrial forms of agriculture. Without it, we would all starve. The issue is this: how do we continue industrial agriculture while minimizing its ecological and environmental impact and reducing its carbon footprint? There are a variety of technologies that must be called on if we are to develop a more sustainable form of agriculture. In a *Scientific American* article, Richard Hamilton (2009) noted that agriculture is a human and not a natural invention, and he perfectly summarized the technological challenge ahead:

> Our challenge is to increase agricultural yields while decreasing the use of fertilizer, water, fossil fuels and other negative environmental inputs. Embracing human ingenuity and innovation seems the most likely path. Plants did not evolve to serve humans, and their sets of genes are incomplete for our purposes. The integral role of modifying genes is obvious to all breeders, though sometimes painfully absent from the public's understanding of how modern agriculture succeeds. All breeding techniques . . . exploit modifications to plant DNA. These modifications can take the form of mistakes or mutations that occur during natural cell division in the wild; the natural but random movement, or "jumping," of DNA sequences from one part of a plant's genome to another; the random genetic changes induced by plant breeders; or the more precise insertion of known gene sequences using biotechnology. In all these cases, plant genes are moved within or across species, creating novel combinations. Hybrid genetics—the combination of different versions of the same gene—has resulted in spectacular yield increases. Largely

as the consequence of using hybrid seed varieties, corn yields in the U.S. have increased more than 500 percent in the past 70 years.

On the other hand, the quest for more efficient and productive farms has led to a heavy reliance on a limited number of plant and animal species. This has made the food supply more vulnerable to disease and resulted in an increase in the use of antibiotics and other medications in food production. According to the Union of Concerned Scientists:

> U.S. agriculture rests on a narrow genetic base. At the beginning of the 1990s, only six varieties of corn accounted for 46 percent of the crop, nine varieties of wheat made up half of the wheat crop, and two types of peas made up 96 percent of the pea crop. Reflecting the global success of fast food, more than half the world's potato acreage is now planted with one variety of potato: the Russet Burbank favored by McDonalds. . . . Decline in the genetic diversity in agriculture is important for a number of reasons. Crops that are very similar to each other in yield and appearance are also similar in their susceptibility to disease.
>
> (UCS 2007)

With the growth of information and communication technology, I think that industrial farming can adopt the more tailored and fine-tuned approach that we see in other forms of manufacturing. Just as the single assembly line of the Ford Motor Company could only produce any color of Model T—as long as it was black—today an auto plant can produced hundreds of varieties of almost custom-made cars. There is every reason to believe that such variety can be developed in agriculture as well.

Up until now, most of the focus of food technology was on improving yield and enhancing profit. While these two goals will always remain, we need to add an emphasis on environmental sustainability to the agricultural equation. Can that be done? In order to answer that question, we must begin by looking at the financial impediments to sustainable agriculture.

In attempting to solve some of these problems, a number of scientists have looked to genetically modified organisms. Some believe that GMOs have the potential to cut back pesticide, herbicide, and fertilizer use while increasing the amount of available food. Forty percent of agricultural loss is caused by weeds, pests, and diseases. Plants can be genetically modified to inherently resist pests and repel weeds without the use of applied pesticides and herbicides. Pesticides and herbicides cause surrounding ecosystem and organism destruction, so reducing our use of them would have significant environmental benefits. However, switching to GMOs might just be replacing one bad thing with another. One study states, "the release of genetically modified organisms into the environment is frequently compared to the introduction of species into a novel environment" (Peterson et al. 2000, 3). While they may reduce ecosystem damage caused by pesticides and herbicides, GMOs cause their own ecosystem damage when their fortuitous genes outcompete the species living in nearby habitats. The UK government's independent wildlife adviser English Wildlife claims, "weed control in these modified crops was more effective and reliable than conventional intensive agriculture . . . this risked further reducing already impoverished farmland wildlife by destroying even more of the weeds it depended on" (BBC News 2004). Until more research is done on their effects on the ecosystem, using GMOs instead of pesticides or herbicides could be a problem instead of a solution.

Because much of the world's arable land is already used for agricultural production, feeding the world's growing population will increasingly mean growing more food within the space that we have. GMOs are one way to accomplish this, but as I previously noted, feeding the starving is less about food quantity and more about access to food. In addition, there are other technologies, such as vertical farms, that could allow us to increase the food supply on the land that we already have without relying on genetic modification.

Finally, scientists and governments have expressed health concerns about how the modified genes in GMOs affect human bodies. The reaction to general modification in the European Union may be instructive. The *Western Daily Press* (2008) reports: "We should not forget that there are no guarantees that GM crops are safe, sustainable or the solution to the problem of hunger . . . more than 70 percent of citizens, Gloucestershire County Council and several governments in the EU have expressed concerns over the negative effects that such crops may have on human health, biodiversity and the environment."

In May 2010, the European Union took a big step to show its disapproval of GMOs by allowing Madeira to officially disallow the growth of any biotech agricultural products in the region. The *New York Times* reported:

> Individual European countries and regions have banned certain genetically modified crops before. Many consumers and farmers in countries like Austria, France and Italy regard the crops as potentially dangerous and likely to contaminate organically produced food. But the case of Madeira represents a significant landmark, because it is the first time the commission, which runs the day-to-day affairs of the European Union, has permitted a country to impose such a sweeping and definitive rejection of the technology.
>
> (Kanter 2010)

GMOs will not be a sustainable solution until scientists, governments, and consumers can feel confident that these technological innovations will not threaten surrounding ecosystems or human health. There is great potential here, but we must be sure that the benefits are greater than the costs.

Financial Issues

Agriculture is a business. In many respects, it is one of the world's oldest forms of business, since the presence of surplus food was the first form of human wealth. In some cases, agriculture is a small business, and in the developing world it can even be a microenterprise. Increasingly, however, it is a massive "agribusiness" that often involves huge, multinational corporations and networks of suppliers, including smaller farms operating under contract to major food companies. In all cases, it is a risky business that depends on weather conditions and the vagaries of local and world food markets.

It is also a business where production (the "growing season") requires an upfront investment of capital for seed and fertilizer, costs that must be carried until revenues are generated at harvest time. Since the price of crops is unpredictable, there are many examples of food growers going bankrupt when food prices drop and farmers not being able to afford the costs of seed and fertilizer for the next planting season (Associated Press 2009).

In addition to the costs of seed and fertilizer, which are annual expenses that must be funded, capital is needed for expenses that can be amortized over a longer period of time. These longer-term expenses include land, buildings, animals, irrigation infrastructure, tractors, other vehicles, and farm equipment that can automate farming functions. The technological base and relative productivity of farming varies widely.

The food business has been radically transformed in the past century. This is the impact of a planet that is now more than half urbanized. While the number of farms in the United States declined through most of the twentieth century, with small family farms being replaced by large industrial farms, over the past decade that trend has been reversed. The U.S. Department of Agriculture reports that "the number of farms in the United States has grown 4 percent and the operators of those farms have become more diverse in the past five years, according to results of the 2007 Census of Agriculture" (Northwestern University Library 2009).

While there has been a revival in the number of farms recently, it followed a century-long decline in the role of agriculture in the American economy. At the start of the twentieth century, about 40 percent of our workers were employed in agriculture. Thirty years later, this was nearly cut in half, and currently a little less than 2 percent of our workforce is involved in food production. In 1930, agriculture was 7.7 percent of the GDP; today it is about 0.7 percent (USDA 2007).

In a 1990 article in *Scientific American*, John P. Reganold discussed the pressure on farmers to preserve soil but also get enough use from that soil to remain in business. He also observed that the high level of small-farmer indebtedness tended to cause them to be risk averse and slow to change farming practices (Reganold 1990).

The growth of industrial farming has provided an increased emphasis on large-scale production and provided some protection from the boom-and-bust cycle that has been characteristic of family farming. At the same time, it has served to increase the pressure for ever-greater rates of return on invested capital. The advantage of food as a magnet for capital is that demand will only increase and will never fade. Food is not a fad, and the demand for nutrition is as sure a consumer item as one can imagine.

While nutrition is not optional, specific foods are (excuse the pun) a matter of taste. Culture, price, and even fashion can influence food trends and agricultural markets. Certain types of foods are favored during good financial times, and others are favored during hard times. Weather and other unpredictable factors influence food productivity and therefore influence the price of capital for agriculture.

Organization and Management

Food is produced, transported, processed, and distributed in a wide variety of ways, whose size and scale varies greatly. Even the largest food producers often make use of networks of farms and firms that vary from multinational corporations to family farms. This contract farming provides flexibility to large-scale food companies and allows them to minimize their exposure to risk.

One of the characteristics of industrial agriculture is that foods are packaged, stored, and shipped over long distances. In all cases, this requires substantial outlays of energy; in some cases, goods are shipped refrigerated or frozen. Vacuum containers and preservatives are common, and technology is used at virtually every step of the process of growing, processing, shipping, storing, and selling food. This has resulted in the development of a number of large and complex multinational food companies (Colcanis 2003).

These companies use the same cutting-edge management practices and technological tools common to any large global corporation. While the reality of the food business is large scale and industrial, there is a great deal of nostalgia and mythology about food production, stemming from the time when a majority of the human population farmed, fished, or hunted for its own food. This period has come to an end. As noted throughout this volume, in 2007, for the first time, a majority of the human population lived in cities (Moreno and Warah 2006). The long era of rural and agrarian lifestyles has been replaced by an urban one that requires massive, interconnected systems of food production and distribution. Some elements of food production, like elements of industrial production, require small organizations that participate in large interorganizational production networks.

The issue of environmental sustainability is one that must be factored into food production, just as it is factored into industrial production. Energy inputs must be analyzed for renewability and carbon footprint. Water use and the ecological impact of fertilizer and pesticides must also be understood. Organizational capacity needs to be developed to both analyze these impacts and reduce them. In some cases, we can expect sustainable practices to arise as a result of market forces that encourage cost reduction. Companies that specialize in working with other organizations to help them reduce energy and water use have already entered the marketplace, and as these costs continue to rise, we will see more such entries.

The capacity of small and more traditionally based farms to adopt new practices and technologies is limited by culture, capacity, and cash. The successes we have seen in transforming traditional farming have been mainly the result of large multinational food companies providing capital, equipment, and training as part of contract farming (Eaton et al. 2001).

In order to ensure that food supplies are appropriate for processing and are able to meet health and quality standards, some large food companies have devoted resources to helping small farmers upgrade their capacity. In certain cases, they have simply bought out these farms, but in other cases, they have worked with small farmers to nurture their capacity and increase their productivity. Over time, I expect to see more of these smaller land holders brought into more integrated food production systems, managed by large multinational food companies. While there are good reasons to fear this trend, I do not see any way for it to be reversed and thus every reason to work to ensure that these relationships do not destroy local agricultural communities.

The Role of Government and Public Policy

Government has a key role to play in developing energy and water efficiency standards and in funding the basic research and development of the science, engineering, and management practices that underpin agricultural sustainability. Government will also need to develop and implement a set of rules, or a regulatory regime, to ensure that sustainable agriculture is required and not simply an optional best practice. The tragedy of the commons is a story of overuse of agricultural resources: the grazing land of the commons. The logic of the market alone does not lead to sustainable agriculture. Rules must be set and enforced as well. In addition, just as America's land grant colleges developed the basic technology of agriculture in the nineteenth and twentieth century (USDA 2009), governments must fund university and think tank–based research on the technology of agricultural sustainability. Those technologies will then need to be provided to the private sector for commercialization and ultimately use by the food industry.

Just as in other areas of sustainability, short-run profits may be the enemy of long-term and sustainable practices. Smaller and weaker organizations may not have the organizational or financial capacity to take the long view. Again, this is where government can play a creative role in making it profitable for these companies to forgo some short-term gains for longer-term benefits. Incentives can be directed to the private sector by using tax laws, the power of government food purchases, and other policy mechanisms to encourage sustainable food practices.

There are several major sustainability issues generated by industrial agriculture. The first relates to the use of nonrenewable resources such as fossil fuels. The second is the contamination of potentially renewable resources such as soil and water by short-sighted agricultural practices designed to drive up crop yields for a few years but that destroy critical agricultural resources. A third issue closely resembles the pollution that comes from nonagricultural industrial facilities. Feedlots and other facilities for raising beef and cattle often have major problems containing and treating animal waste products before releasing them into local waterways. The potential for recycling animal waste as fertilizer is great, but extraordinary care and skill is needed to ensure that this waste is safely reprocessed for use. Inadequate capital investment and poor management have sometimes resulted in substantial levels of environmental damage near agricultural facilities (Pew Charitable Trusts 2008).

In part, the role of government is to create a set of environmental rules that remove some of the incentives to pollute. However, because of the global nature of industrial agriculture, it is important to work to ensure that these rules are applicable throughout the world, to prevent the export of both pollution and manufacturing. While pollution control can add costs to agricultural production and requires government rules and enforcement, energy, water, and resource efficiency can reduce costs and requires a different approach.

Resource efficiency requires capital for equipment and training for best management practices. Government can use tax rules and loan guarantees and can fund basic research into technology that can have the effect of reducing the cost and risk for private companies willing to take the steps needed to improve their sustainability.

Another role for government policy is in the area of the laws governing the import and export of food and on food safety. Since food is grown all over the developing world for consumption everywhere, limits on the use of toxins in food manufacturing are growing and will continue to grow. However, government subsidies and trade limits have the effect of distorting agricultural markets and can have unpredictable effects. When governments formulate and implement food policies, it is important that they assess the impact of those policies internationally as well as domestically. A sophisticated use of government policy can enhance agricultural sustainability. Unfortunately, as in the United States, agricultural policy is often defined by interest groups, economic forces, and ideological mindsets that provide narrow benefits at significant societal cost. There is a long history of government intervention in agricultural markets throughout the world. While some of these policies have helped farmers survive the downturns of volatile economic conditions, many persist after their useful life has ended. Bringing sustainability considerations into this interest-packed policy arena is difficult and will take a determined effort over a long period of time.

Conclusions

The sustainability of food supply is a fundamental element of the field of sustainability management. Food is not an option, and nutrition and health are closely interconnected. Like water and air, food is a necessity for human survival. Because of the increased urbanization of our settlements, we live farther and farther from the sources of food, and as time goes on we all are moving further and further from our agrarian roots. We need to learn how to feed ourselves without destroying the land and water resources required for agriculture. This is one of the critical sustainability challenges of the twenty-first century.

CASE STUDY: LOCAL AND ORGANIC FOOD

Growing population, drought, the use of biofuels, and increased consumption have contributed to a growing world food crisis. The most vulnerable among us here in the United States and poor people around the world are most affected by rising food prices and shortages. At the same time, we also see a growing awareness of where our food comes from and the environmental impact of food production. One result of that has been the increased use of organic and locally grown foods. The movement toward using these foods is easy to see in our supermarkets and in the sidewalk greenmarkets located in many communities.

Organic food sales grow every year, increasing by over 20 percent in 2003 alone, according to the Organic Trade Association. Just over 6 percent of all produce sales now fall into the organic category, up from 2.5 percent a decade ago (OTA 2010). Organic food benefits the environment, local communities, and public health. It does not rely on synthetic or petroleum-based pesticides or fertilizers, resulting in less water and soil contamination due to runoff.

It also preserves local open land, which is rapidly disappearing. According to American Farmland Trust, "New York lost 127,000 acres of farmlands between 1997 and 2002—an average of 70 acres of farmland a day" (Dunlea et al. 2005, 17). Buying organic from local farmers' markets also reduces your carbon footprint. On average, food travels 1,300 miles before reaching your plate, a process requiring a large amount of energy. According to the Council on the Environment of New York City (CENY), "it takes 435 fossil fuel calories to transport a 5 calorie strawberry from California to New York" (Dunlea et al. 2005).

CENY runs the city's greenmarkets and has outlined some additional reasons why farmers' markets are good for New York.

Food Security: "Greenmarket participates in the NYS Farmers Market Nutrition Program, providing food to families at nutritional risk. In 2005, almost 250,000 such households redeemed vouchers worth $3 million for locally grown fresh fruits and vegetables at NYC farmers markets" (Dunlea et al. 2005). In 2005, Greenmarket donated over 300,000 pounds of food to City Harvest.

Improvement of Neighborhood Economies: "In peak season, the Union Square Greenmarket draws 60,000 shoppers a day; in a recent survey, 82% cited Greenmarket as the primary reason for their visit, and 60% spent up to $50 in area businesses" (Dunlea et al. 2005).

Biodiversity Preservation: "Greenmarket farmers grow thousands of varieties of fruits and vegetables, including over 100 varieties each of apples

(continued)

CASE STUDY: LOCAL AND ORGANIC FOOD (*continued*)

and tomatoes. In contrast, industrial agribusiness cultivates high-yield hybrids bred for fast maturation and thick skins to withstand mechanical harvest and transport. The UN Food and Agriculture Organization estimates that more than 75% of agricultural genetic diversity was lost in the 20th century. Small, biodiverse farms preserve our food heritage" (Dunlea et al. 2005).

Goods labeled "Certified Organic" are strictly regulated, because the word "organic" is central to the certification and marketing process. With the exception of smaller growers who sell less than $5,000 in goods per year, organic farmers are inspected at least once annually to ensure compliance with National Organic Program standards on production and processing. Some criticize this process for not being adequately inclusive. Organizations such as Certified Naturally Grown offer a "non-profit alternative eco-labeling program for small farms that grow using USDA Organic methods but are not a part of the USDA Certified Organic program" (Organic Seeds Trust 2010).

New York law makes it the state's policy to encourage the creation and use of farmers' markets in promoting agriculture. The law states: "The legislature hereby finds and declares that farmers' markets provide a vital and highly effective marketing mechanism for thousands of New York farmers, improve the access of consumers and wholesalers to New York farm products, and contribute to the economic revitalization of the areas in which the markets are located" (Findlaw.com 2010).

The volume of food needed to feed a city of this size makes large-scale food manufacturing a necessity. A May 2005 survey by the New York State Department of Agriculture and Markets reported that New York represents a $30 billion per year market for food. The market demand for locally grown and processed products amounts to more than $866 million per year (Organic Seeds Trust 2010). In addition, in New York City and other urban environments, community gardens can play a big role in diversifying both the city experience and the city dweller's diet.

CASE STUDY: COMMUNITY GARDENS IN NEW YORK

While New York is nearly completely built up, there are places within the city where there is enough land to grow some crops. We are certainly surrounded by concrete and asphalt, but the natural world is never far away in New York City. From small plots to multiacre urban farms, New York City's community gardens turn abandoned lots into urban oases, feed city residents, and provide community spaces for birthdays, barbeques, and informal get-togethers.

In addition to benefits such as fostering community and offering green spaces in neighborhoods lacking sufficient parkland, community gardens also have a positive impact on the environment. Unpaved garden surfaces absorb rainwater and reduce stress on the city's sewage system, and many gardens partner with schools to provide outdoor classrooms for ecology and biology lessons.

Community gardens help cool the city and reduce the urban heat-island effect, caused when the city's dark surfaces trap heat, making it hotter than surrounding areas. Green spaces offered by community gardens can even reduce the amount of energy used to cool buildings. According to a study by the U.S. EPA, "trees provide shade, cool the interior [of buildings], and reduce air conditioning demand" (U.S. EPA 2003, 1).

While most of our food travels over a thousand miles by the time it reaches grocery shelves (NSAIS 2010), community gardens are a source of fresh, affordable produce for city residents that can be transported to the dinner table without the use of fossil fuels.

As Jacquie Berger, executive director of Just Food, points out, "as food prices rise, people are trying to figure out how they can get food more affordably. With community gardens, people can get together to grow their own food, which builds community, saves money and shrinks their carbon footprint all at the same time."

Just Food currently works with more than thirty-five community gardens to help the gardeners grow and sell their produce (Just Food 2010a). Through its Community Food Education Program, Just Food pays experienced gardeners to host workshops all over New York City on such topics as seed starting, raising chickens, building raised beds, season extension, food preservation, and even making baby food (Just Food 2010b). "The Brooklyn Rescue Mission grew 7,000 pounds of produce last year at their garden in central Brooklyn," says Berger. "That year they were growing primarily for their food pantry, but they had so much that this year we are helping them start a farmers' market as well."

GreenThumb, a division of the NYC Parks Department, is the largest urban gardening program in the country. They work with more than six hundred gardens, providing technical assistance and materials. While some gardens can be pretty large, ornamental gardens as narrow as fifteen feet can provide shady areas and benches where community residents gather (NYC DPR 2010).

(*continued*)

CASE STUDY: COMMUNITY GARDENS IN NEW YORK (*continued*)

Added Value, a nonprofit in the Red Hook neighborhood of Brooklyn, created an urban farm by transforming 2.75 acres of city asphalt into fertile land. The farm is now the site of a youth employment program, a farmers' market featuring produce from Added Value and other regional farmers, and a large-scale composting operation that accepts waste from area businesses. Neighborhood restaurants proudly hang signs boasting that their menus feature Added Value's produce (Added Value 2009).

Other groups have used gardens as a springboard for organizing around a diverse variety of community issues. La Familia Verde, a coalition of gardens and organizations in the Bronx, has partnered with community-based organizations to organize voter registration drives, health fairs, and a farmers' market (Living Memorials Project 2003).

New York City's urban gardens are rare and threatened treasures. The insatiable demand for NYC real estate puts these gardens under constant pressure. The "More Gardens!" coalition joins with community gardens facing removal in order to fight for their preservation. Using strategies such as camping out—sometimes for months—in gardens slated for demolition, enlisting the support of local and state politicians, and even filing a lawsuit against the city, the coalition has helped save more than four hundred gardens from development (More Gardens! 2010).

Community gardens eliminate the expenditure of energy used to transport food; provide cool, green spaces during the sweltering summer months; and give New Yorkers the most local food possible. When coupled with the local food produced just north and west of the city and sold at greenmarkets, they provide an important and affordable alternative to industrial farming. While a city as large as New York will always need mass agriculture, every piece of locally grown and used food is a small step toward sustainability.

Chapter 6

Sustainable Cities

Definition and Technical Challenges

What is a sustainable city? A city, by definition, is a place that is not designed to be fully in harmony with natural ecosystems but is instead dominated by human activities: our homes, businesses, and institutions. When I think of a city that could be termed sustainable, I think of a place that uses as few nonrenewable resources as possible and has the least possible impact on the ecosystems outside of the city's boundaries. That concern for external impact requires a concern for the impact on the already compromised ecosystems within the city, although there are limits to ecosystem preservation options within cities.

What is the accepted definition of an urban area? There are a variety of definitions in use. The U.S. Census Bureau (2009) defines a metropolitan urban area as one with

> a core urban area of 50,000 or more population, and a micro area contains an urban core of at least 10,000 (but less than 50,000) population. Each metro or micro area consists of one or more counties and includes the counties containing the core urban area, as well as any adjacent counties that have a high degree of social and economic integration (as measured by commuting to work) with the urban core.

The World Resources Institute has summarized some of the definitional complexities with the term "urban." They note that there are over 3.7 billion people living in the world's urban areas. They then observe that urban areas can vary from the size of megacities to small towns and that

> all cities are urban areas, but not all urban areas are cities. "Urban" is a statistical concept defined by a country's government. A city, on the other hand, is more than just large numbers of people living in close proximity to one another; it is a complex political, economic, and social entity. Cities around the world symbolize their nation's identity and political strength. Cities are also centers of economic production, religion, learning, and culture.
>
> Because each country sets its own definition of "urban," there is a bewildering array of definitions around the world. . . . In many countries, the definition is based on a threshold number of inhabitants. . . . This threshold ranges from a few hundred, as in Peru and Uganda, to more than 10,000, as in Italy and Senegal. . . . Other governments base their definition on a combination of criteria, such as population density, political function, or predominant activity of the region.
>
> (WRI 2010)

Both definitions are clear that urban areas are concentrations of humans and their activities. Concentrations provide both opportunities for cost-effective infrastructure that takes advantage of economies of scale and causes problems due to the impact of this population on humans and on the ecosystems that surround them. However, it is one thing to create such settlements; it is another thing to sustain them over time (Cohen 2006, 10). This chapter will focus on the issue of urban sustainability.

To conceptualize urban sustainability, it is helpful to deconstruct the systems that make it possible for us to live in concentrated areas of land. This includes:

- The way we use land and its impact on the air and water that passes through
- Materials used to construct and maintain our built environment

- Transport systems that allow us to move around our cities
- Energy that powers our built environment and transport systems
- The way we use water and what we do with it once it has become a waste product
- The way we transport, store, and dispose of nontoxic and toxic solid and liquid wastes
- The quality of our air
- Economic systems that provide us with the means to earn a livelihood
- The cultural, educational, recreational, health, and entertainment institutions that provide many of the positive benefits that we derive from living in cities

These systems are the building blocks of urban areas and are the elements of cities that must be sustained if cities are to be sustainable. Notice that many of the resources discussed in earlier chapters are now integrated in a physical space and discussed in this chapter as elements of an overarching system of human settlement. After discussing these elements of sustainable cities, I will then turn to a discussion of the financial, management, and public policy challenges to urban sustainability.

Sustainable Land Use

The history of the United States in many respects has been a history of the use of land. The concept of land ownership is important to understand. In some cultures, such as some Native American tribes, the concept of land ownership was unknown. In America, the rights attached to land ownership can be quite comprehensive, although in all cases, government retains the right to seize land with appropriate compensation and for a public purpose. This is the right of eminent domain, which is always reserved to government in the United States. As European settlers moved west across the continent, slaves came from Africa, and Asian settlers settled both on the west and east coasts, early surplus wealth was generated by the land's productive bounty. In the mid-twentieth century, the ideal of land ownership coupled with the impact of the auto created the land development pattern that came to be called suburban sprawl.

As development spread throughout the countryside, many cities deteriorated, and the costs associated with our urban infrastructure increased, as new facilities were needed while existing infrastructure needed to be maintained. The same roads, water and sewage systems, power, and other facilities in place in cities also had to be built in rural areas to ensure that the new suburban residents enjoyed the same services they had grown accustomed to in cities. The irony was that while early suburbanites enjoyed lower congestion and more bucolic environments, in less than a generation, these formerly rural areas became the spread-out, suburban-style cities of the late twentieth century. Many of the problems that people left the city to escape—especially traffic, crime, and pollution—just followed these folks to these new urban settlements.

Much of the United States' projected population growth for the twenty-first century could be accommodated by the resettlement of old cities. We are starting to see some of that as some old cities start to revive in the northeastern United States. However, in much of the world, new cities and settlements will need to be built on land that is not now developed. In both cases, development of these settlements could adhere to the principles of development with the least possible environmental impact. This requires great care in designing human settlements.

This is not a new idea. Many cultures over time have paid great attention to development that maintains ecosystems and their productivity. In 1969, Ian McHarg wrote his classic, *Design with Nature*, which popularized the idea of low-impact development. In the days before low-cost computing and sophisticated geographic information systems, his creative use of overlay maps helped identify locations that could be used for development with the least possible impact. McHarg paid particular attention to drainage and the maintenance of fragile ecosystems of particular importance to human well-being.

By concentrating human settlements, managing its waste flows, and developing renewable sources of energy and material flows, it is possible to retain the productive capacity of the planet while accommodating seven to ten billion people. Well-designed land use reduces stress on ecological systems and provides humans with the benefits of technology and civilization. For this to be accomplished, the free market in land use and sale must be regulated through effective and enlightened zoning and government incentives. This will be discussed further in the section below on the role of government and public policy.

When we think about sustainable urban land use, it is important to remember that some of the uses of land take place underground. Subways, water and waste pipes, electrical lines, gas pipes, and even roadways are buried beneath the surface. It is very expensive and disruptive to bury or build infrastructure underground, and it can be very difficult to maintain this infrastructure because of corrosion and water damage. Still, there are a number of practical advantages in using land below the surface. In the case of mass transit, it allows for faster transport within a city. Electrical power lines that could be disrupted by weather are more reliable when they are underground. Underground construction provides some of the land efficiency gained by skyscrapers, without blocking light. Of course, in cities with aging infrastructure, water main breaks and other underground accidents can easily cause massive disruptions.

Sustainable Materials to Build and Maintain Urban Settlements

Modern construction uses a wide variety of materials, ranging from plexiglass to concrete to marble. A growing number of builders and architects are paying serious attention to the sustainability of their building's systems and materials. Lynn M. Froeschle, a San Diego–based architect, has published an excellent article on assessing green building materials. When she analyzes the sustainability of building materials, some of her criteria include low toxicity, minimal emissions, recycled content and suitability for recycling, resource efficiency, reusability, materials from sustainable sources, durability, energy efficiency, water conserving, and the acquisition of the materials from local sources (Froeschle 1999).

Froeschle discusses the management issues involved in using sustainable building materials, noting the difficulty in finding builders with experience in sustainable construction. In the second decade of the twenty-first century, this situation has improved, but the potential for a gap between green design and real-world implementation remains. It's not hard to imagine that some construction managers would consider sustainability issues unimportant or even frivolous. Their desire to cut corners on the job should never be underestimated.

Nevertheless, cost issues and the marketing advantages of sustainability have dramatically increased the availability of contractor experience in this type of construction.

The use of sustainable construction materials should also be extended to the materials needed to restore and maintain structures. Even the most sustainable building can end up wasting energy and water if it is not maintained or operated properly. Like an auto that is allowed to go without a tune-up, a structure can end up running poorly if it is neglected. In this respect, there is a close connection between sustainability and effective and constant attention to a building's operation and maintenance.

Sustainable Transportation

Cities are collections of complex economic, political, social, and ecological systems. The economic and social structures of urban areas are both reflections and causes of their transport infrastructure. In the United States, the interstate highway system and our massive investment in roads reinforced a suburban development pattern. This ultimately has resulted in widespread use of personal rather than mass transportation. The United States has 260 million autos, and nearly all of our transportation is via automobile. And even though cities such as New York, Washington, Boston, and Chicago have well-functioning mass transit systems, with the exception of New York, people in these cities travel more miles by auto than by any other means (VTPI 2010).

Given the commitment to a dispersed pattern of land use in the United States, there are clear limits to the usefulness of mass transit as a strategy for sustainable transportation. Once sprawl development takes place, it becomes very expensive for cities to purchase the land needed for new subway systems. An interesting low-cost alternative is the Red Line bus system that Bogota, Colombia, built. The system uses buses and fixed above-ground stations along with timed traffic lights to create a trolley-style mass transit system. Bogota is as spread out as any city in the United States but has developed an imaginative, low-cost form of mass transit (Rosenthal 2009).

The economic and social need for some concentration of activity requires that mass transit address issues of traffic congestion. Scarce space for roads, vehicles, and parking means that some portion of a city's population must transport themselves without the use of a private automobile. The use of strategies such as congestion pricing can help a city balance personal transit and mass transit by incentivizing the use of more efficient modes of transportation.

In addition to improving mass transit, personal transit must be made more sustainable. An auto must be developed that does not rely on a fossil fuel–burning internal combustion engine. Hybrid autos are a step in that direction, and electric autos are an even greater step. However, just as refueling infrastructure developed for autos running on gasoline, so too must facilities be developed to recharge electric autos. In 2009 and 2010, we saw the momentum for this infrastructure begin to build. In a *New York Times* article published on February 14, 2010, Todd Woody and Clifford Krauss (2010) wrote that "the San Francisco building code will soon be revised to require that new structures be wired for car chargers. Across the street from City Hall, some drivers are already plugging converted hybrids into a row of charging stations." They observed that the technology of electric cars also depended on the development of convenient and inexpensive recharging infrastructure:

> Much of the attention on electric cars has been on the vehicles' design, cost and performance. But success or failure could turn on more mundane matters, like the time it takes car buyers to navigate a municipal bureaucracy to have charging stations installed in their homes. When the president of the California Public Utilities Commission, Michael R. Peevey, leased an electric Mini Cooper, he said, it took six weeks of visits by installers and inspectors before he could plug in his new car at home.

Intracity transportation of people and materials must also be considered an element of sustainable cities. If cities are supplied with food and goods, people from other places will come to visit or conduct business. The systems that bring people and materials to and from cities must be developed to minimize resource use and environmental impacts.

Maintaining mass transit systems—and all other transportation infrastructure—is expensive and complicated. We will see this in the following case study, which is based on a piece I wrote for the *Huffington Post* in early 2010.

CASE STUDY: SUSTAINABLE FINANCING FOR MASS TRANSIT IN NEW YORK

One of the victims of the economic downturn in New York is state support for mass transit. Unfortunately, this is not simply a result of the recent decline in state tax revenues but rather a long-term trend that was exacerbated by overborrowing for mass transit during the Pataki era.

One of the key elements of the New York City metro area's dynamism is its mass transit system. Destroy that critical infrastructure, and you begin to destroy the entire region's economy. New York's high population density requires efficient mass transit. The region's businesses need a well-functioning mass transit system to bring in workers and customers. Our environment is better served by mass transit than by personal transit, and our patterns of land use development have largely followed mass transit routes. The economic importance of New York's mass transit system is an example of the close relationship between economic development and environmental sustainability. Its presence helps make New York City the most energy-efficient city in the country (Hagan 2009).

Mass transit in New York has always suffered from financial difficulties. The original New York City subway system built at the start of the twentieth century was operated by private but heavily regulated franchisees. The insistence on the nickel fare and the Great Depression ultimately resulted in a public takeover of the system in 1940. The post–World War II era saw the removal of many of the remaining elevated lines, including the Third Avenue line on the East Side, which was demolished to make way for the still unfinished Second Avenue subway. In the 1960s and 1970s, capital disinvestment caused the system's near collapse. "By the early 1980s a third of the fleet was typically out of service during the morning rush hours, cars broke down or caught fire, trains derailed on hazardous track, and graffiti covered virtually every car. In 1982 the MTA began to rehabilitate the subways through a series of five-year Capital Programs, the largest public transportation rebuilding effort in national history. Over $39 billion has been invested since the program began" (Melwani 2010).

During the 1980s and 1990s, the city's subway system came back from the brink of ruin because of the effective leadership of people including Richard Ravitch, the former head of the MTA; the former governors Hugh Carey and Mario Cuomo; and the mayors Koch, Dinkins, and Giuliani. These leaders instituted the range of tolls and taxes now used to subsidize mass transit. Unfortunately, under Governor Pataki, the mass transit capital subsidy was sharply reduced, forcing the MTA to use more and more of their budget to pay debt service on transit bonds. Even worse, during the 2010 $6.8 billion state budget crisis, Governor Patterson has further reduced the state's subsidy for the MTA, adding to the transit agency's $400 million budget gap.

The MTA's response to Patterson's proposed funding reduction was the same one we often see when school board budgets are voted down by local communities. It's what I call the "football team gambit": Cut the most visible and popular expenditures and hope to stimulate a public outcry that results in budget restoration. In this case, the MTA started by announcing that they would cut free rides for schoolchildren. For a variety of reasons, there are very few yellow school buses in New York. It is far more cost effective for public school kids to use the same public buses and subways used by other commuters. The MTA has long subsidized these fares and estimates that the subsidy costs them $170 million per year (Grynbaum 2009).

The next highly visible cut proposed by the MTA is to put out of service some buses and trains now running late at night and on weekends (Grynbaum 2009). Twenty-four/seven mass transit has long been considered an essential right in the city that never sleeps. New Yorkers know that subways that close at night are for small towns like Washington, D.C., where dinner starts at 6 P.M., or Miami, where early birds dine at 4:30 in the afternoon.

From the perspective of sustainability and energy efficiency, we want to do everything we can to make mass transit safe, fast, comfortable, and cheap. But in order for it to compete with the private auto, mass transit must be subsidized, and some of those subsidies should come from private autos, to help raise the cost of driving and reduce the cost of riding.

Of course, all of these service cuts are merely proposals. Now, we begin the political Kabuki dance known as the New York State budget process. You can bet that New York's schoolchildren will not end up paying for their transportation to and from school. Many of these proposed service cuts will be miraculously restored. Some of the cuts will remain. The most endangered part of the budget is the long-term capital budget. The crucial dollars that are needed to keep the system in good repair and to continue its expansion will be cut. We've been down that road before, and, as everyone knows, deferred maintenance increases long-run costs even while providing short-term budget relief.

Maybe now the New York State legislature will take another look at congestion pricing for New York City's midtown and downtown business districts. A charge on vehicles entering Manhattan south of Fifty-ninth Street could generate the funds now being cut and also reduce traffic, energy use, and air pollution. Perhaps the stark choices ahead for New York will result in some creative and imaginative policymaking after all. Of course, we could always follow California's lead and simply close the store.

Sustainable Urban Energy

The earlier chapter on energy discussed the complexities of energy sustainability but did not discuss the generation and distribution of energy within cities. Urban energy systems rely on centralized and typically fossil fuel–based power plants and power transmission lines owned by government-regulated monopolies. With a centralized generation system and a transmission system built on a grid, cities provide concentrations of customers who can increase the efficiency of energy infrastructure. Similarly, the concept of distributed power generation makes it possible for consumers to not only draw power from the grid but contribute to the grid by generating power in small increments at home or in small, decentralized, renewable facilities such as windmills.

For a city to function, it must have plentiful and reliable sources of energy: electricity for private homes, transport, and businesses as well as public uses such as streetlights and other fuels for climate control and transportation. In the near future, we expect that more and more transportation will be powered by electricity. Cities must plan to ensure that their energy supply is adequate. This involves sufficient generation and transmission capacity and increasing the amount of attention paid to energy efficiency.

New York City's sustainability plan, PLANYC2030, delineates eleven specific energy initiatives that provide a good sense of what sustainable urban energy is all about:

1. Establish a New York City Energy Planning Board
- Work with the state and utilities to centralize planning for the city's supply and demand initiatives
2. Reduce energy consumption by city government
- Commit 10 percent of the city's annual energy bill to fund energy-saving investments in city operations
3. Strengthen energy codes in New York City
- Strengthen our energy and building codes to support our energy efficiency strategies and other environmental goals
4. Create an energy efficiency authority for New York City
- Create the New York City Energy Efficiency Authority (NYCEEA), which will be responsible for reaching the city's demand-reduction targets

5. Prioritize five key areas for targeted incentives
- Use a series of mandates, challenges, and incentives to reduce demand among the city's largest energy consumers.

6. Expand the Promote Peak Load Management program
- Expand participation in Peak Load Management programs through smart meters
- Support the expansion of real-time pricing across the city

7. Launch an energy awareness and training campaign
- Increase the impact of our energy efficiency efforts through a coordinated energy education, awareness, and training campaign

8. Facilitate repowering and construction of power plants
- Facilitate the construction of 2,000 to 3,000 MW of supply capacity by repowering old plants, constructing new ones, and building dedicated transmission lines

9. Expand Clean Distributed Generation ("Clean DG")
- Increase the amount of Clean DG by 800 MW
- Promote opportunities to develop district energy at appropriate sites in New York City

10. Support the expansion of gas infrastructure
- Support critical expansions to the city's natural gas infrastructure

11. Foster the market for renewable energy
- Create a property tax abatement for solar panel installations
- Study the cost effectiveness of solar electricity when evaluated on a real-time pricing scenario
- Support the construction of the city's first carbon-neutral building, primarily powered by solar electricity
- Increase use of solar energy in city buildings through creative financing
- Work with the state to eliminate barriers to increasing the use of solar energy in the city
- Pilot one or more technologies for producing energy from solid waste
- End methane emissions from sewage treatment plants and expand the productive use of digester gas
- Study the expansion of gas capture and energy production from existing landfills

12. Accelerate reliability improvements to the city's grid
- Advocate for Con Edison to implement recommendations from the city's report on the northwestern Queens power outages

13. Facilitate grid repairs through improved coordination and joint bidding
 - Pursue the passage of joint bidding legislation
 - Ensure that adequate pier facilities are available to Con Edison to offload transformers and other equipment
14. Support Con Edison's efforts to modernize the grid

(Office of the Mayor, City of New York 2009)

The overall goal is to ensure an adequate and reliable energy supply, and the technique used to accomplish this goal is to reduce energy waste and increase the use of renewable sources. While New York is the largest local government in the United States and might not be an excellent model, the range of energy initiatives it is undertaking is quite comprehensive and might be thought of as a menu of options for smaller urban areas. The initiatives focus on energy efficiency and renewable energy as well as on building institutions to plan and implement energy programs.

To a considerable degree, a city's energy use will be determined by actions taking place beyond its borders. Urban areas will be influenced by the actions of national governments and by the technologies developed and commercialized by the private sector. However, the degree to which a city takes advantage of resources and new technologies generated externally will strongly influence the city's competitiveness in the global marketplace. Local government officials are becoming increasingly aware of this. As a result, large numbers of them have begun sustainability initiatives like New York's PlaNYC. At the December 2009 Copenhagen climate talks, over one hundred mayors from around the world met to discuss the practical problems of local-level climate and energy policy. It is clear that the issue of climate and energy has moved to the center of the urban sustainability agenda.

The key in planning for urban energy is to ensure that there is always sufficient energy available for use. It should be delivered at the lowest possible cost and with the least environmental impact possible. Fuel supplies and transmission infrastructure are as important to these goals as the actual generation facility. Sustainable energy in urban areas should take advantage of economies and efficiencies of scale but should not neglect the possibilities of decentralized energy sources. Future energy technologies might reduce homeowner dependence on the grid. Urban building codes and other regulations will need to be mindful of those trends and accommodate them when possible.

Interruptions in energy supplies can have a devastating impact on urban life. Transportation, communication, production, food and water delivery, and health care depend on the constant flow of energy. Given the interdependence of modern cities, people, and production systems, the impact of frequent power outages or even brownouts can be massive and unpredictable. This factor often leads to the construction and operation of redundant or wasteful systems designed to avoid failure. However, the need for safety measures is undeniable. A sustainable urban energy system must assume this precaution as a design parameter and then work to maximize the system's efficiency and cost effectiveness while minimizing its environmental impact.

Sustainable Water and Sewage Treatment

While the earlier chapter on water noted that most of the world's freshwater was devoted to agricultural use, nearly all of the water supplied to cities is for human consumption. In an article in *The Environmental Impact Assessment Review*, Daniel Hellström, Ulf Jeppsson, and Erik Kärrman presented "a framework for systems analysis of sustainable urban water management." The authors observe that "urban water and wastewater systems should—without harming the environment—provide clean water for a variety of uses, remove wastewater from users to prevent unhygienic conditions, and remove storm water to avoid damage from flooding" (Hellström et al. 2000, 311–312).

They then develop a framework and a set of criteria for assessing the sustainability of an urban water system. They explicitly state that the analysis must go beyond simple economic costs and benefits to factor in social, cultural, and environmental impacts. They note: "the proposed set of sustainability criteria have been divided into five main categories: (1) health and hygiene criteria, (2) social-cultural criteria, (3) environmental criteria, (4) economic criteria, and (5) functional and technical criteria" (Hellström et al. 2000, 315).

Sustainable sources of freshwater must have the capacity to be resupplied with clean, fresh water. This means that deep groundwater sources cannot be relied on, since those sources are not currently being replenished (McReynolds et al. 2005). Unless natural ecosystems provide sufficient filtration of the water supply, artificial water filtration systems must be constructed and maintained. These systems require extensive capital to construct and a great deal of energy to operate and maintain.

Once water has been used, it becomes a waste product. When combined with storm water, freshwater must be treated and cleaned before it is returned to the waterways. The sewage treatment process is itself energy intensive and produces a sludge that must be further treated or landfilled. The pipes that carry water to and wastewater from our buildings must be buried underground, and they must be maintained over time.

As vital as energy and transportation infrastructure are to a city, in many respects the water system is even more important. Human beings can go without mobility and energy much longer than they can go without water. A sewage system that ceases to function would quickly create a public health emergency. The technology of waste treatment has advanced dramatically over the past seventy-five years, and the effluent discharged by wastewater treatment plants continues to improve in quality. Evidence of these advances can be seen when treated sewage water is used to water lawns in parks and fairways on golf courses.

While the unpredicted interruption of supply is less of a problem than it is with urban energy systems, even the best-designed water systems suffer from periodic droughts. Just as with electricity, however, interruptions of a city's water system can have a cascading and unpredictable effect on a city's social, cultural, and economic life. As noted above, the human ability to survive without water is far lower than our ability to survive without electricity. Of course, just as portable generators can provide energy during blackouts, bottled water and alternative means of maintaining the water supply can be implemented during emergencies. However, the image and reputation of a city can be irreparably harmed if interruptions in water supply are frequent. Water of insufficient quality can have a similar impact. It is not enough to have plenty of liquid coming out of the pipe. It needs to be usable for bathing, drinking, and cooking.

Sustainable Solid and Toxic Waste Management

In addition to the sewage waste just mentioned, the humans and animals that live in cities produce a great deal of solid waste that is not flushed down the toilet. Every day in 2008, each person in the United States generated an average of 4.5 pounds of solid waste. Each day, the 304 million or so people who live in the United States generate about 1,368,270,000 pounds of waste. In 1960, we generated only 2.68 pounds of waste a day. With about 180 million people in the country, our total solid waste load was 482,343,720 pounds (U.S. EPA 2008). The good news is that our waste per capita peaked at 4.65 pounds per day in 2000, and of the 4.5 pounds of waste we generate per day, we recover 1.5 pounds in materials and burn a little more than .5 pounds per day in waste-to-energy incinerators. That leaves 2.43 pounds per person per day sent to our landfills. This brings us back to 1960 per capita disposal rates, although we now have about 125 million more people (U.S. EPA 2008).

We are getting better at recycling and reusing materials, but we need to improve our management and technology. Americans generate lots of garbage, and its management and use is one of the fundamental barriers to achieving urban sustainability. Moving garbage from its point of generation to a point of reuse or disposal is a complex and energy-intensive operation.

New York City perfectly exemplifies the dilemma of solid waste management. Until 2000, most of New York's waste was buried in landfills within the city's borders. Since the start of the century, 100 percent of the city's collected waste has been transported out of the city to landfills and incinerators. The following case study is based on a piece I wrote for the *New York Observer* summarizing the waste problem in New York City and suggesting possible sustainable solutions.

CASE STUDY: WASTED—NEW YORK CITY'S GIANT GARBAGE PROBLEM

New York City's eight million residents and millions of businesses, construction projects, and visitors generate as much as 36,200 tons of garbage every day. The city's Department of Sanitation handles nearly 13,000 tons per day of waste generated by residents, public agencies, and nonprofit corporations; private carting companies handle the remainder. During the twentieth century, the city relied on a number of landfills for garbage disposal. Then, in December 2001, the city's last garbage dump, Fresh Kills

(continued)

CASE STUDY: WASTED—NEW YORK CITY'S GIANT GARBAGE PROBLEM (*continued*)

Landfill in Staten Island, closed. In response, we adopted a twenty-year plan for exporting waste.

The city's annual bill for collecting and disposing of residential trash jumped from about $658 million in 2000 to about $1.25 billion in 2008. The cost of disposal has grown from $300 million in 2005 to about $400 million in 2008 (Niblack 2007). While some of that increase is because of inflation, most of it is because of the higher cost of transporting and landfilling garbage out of state. The city's long-term plan is to reduce costs by recycling more, reducing waste, and building a waterfront waste transfer system less dependent on trucks and able to use containers to ship garbage by barge and train further away to cheaper dumpsites.

It is hard to imagine a more environmentally damaging waste management system than the one we have in New York City. Actually, it's not so hard to imagine, if you look back and remember when we dumped our garbage into the ocean or used incinerators in the basements of apartment buildings to burn garbage at night (Walsh et al. 2001). Yet today, we collect garbage with trucks that use high-polluting diesel fuel and then dump that garbage onto the floors of waste transfer stations that are typically located in poor neighborhoods. We then scoop the garbage up off the floor and load it onto large trucks that also burn highly polluting diesel fuel and ship it to landfills and waste-to-energy incinerators located away from New York City (*Gotham Gazette* 2007).

While we own our entire water system, our waste system leaves us at the mercy of the private marketplace and the whims of Congress and other states. The current system of waste export leaves the city vulnerable over the long run. For example, it is harder to site landfills in this region than it used to be. Political opposition to landfilling is growing in many communities that already are home to dump sites. Bills are regularly brought before Congress that would authorize local governments, state governments, and governors to restrict or prohibit the receipt of out-of-state waste.

Though the passage of such bills is far from certain, the possibility of passage over the next twenty years is substantial enough to warrant concern. Similarly, stricter regulations on new landfills by federal and state environmental protection agencies could increase the cost of new landfills and limit future landfill capacity. Finally, landfill operators will certainly raise prices over time, and state and municipal governments will likely enact taxes on waste disposal.

Why do New Yorkers create so much garbage? Well, there are a lot of us, and New Yorkers are busy people—we toss garbage causally and we don't like to sort it. We prefer not to think about garbage or where it will end up. I think we have this fantasy that those green plastic mounds of garbage bags on the street are magically transported to some mythical solid waste heaven.

New York's elected leaders know that waste is a no-win issue. As long as the cost increases of exporting waste are gradual, it is unlikely that enough political noise will be generated to induce a sitting mayor to rethink waste export. And no mayor in his or her right mind would try to build a waste incinerator or landfill in or near the city.

Still, the technology of waste incineration has advanced dramatically since we stopped using those horrible apartment incinerators in the 1960s. In Japan, 70 percent of all waste is burned, generating electricity in the process. While incineration pollutes the air, it is less polluting than transporting waste in diesel-fueled trucks to leaking out-of-state landfills.

What is the solution? In 2003, I proposed barging our garbage to waste-to-energy plants located in some of the economically distressed cities along the Hudson River. This could provide jobs and cheaper power to towns that could really use them. While I still like that idea, no one else did.

The next idea I'd like to propose is to develop community-based waste management facilities. Perhaps smaller-scale waste-to-energy plants coupled with recycling facilities and anaerobic digesters (a form of automated compost facility) could be located in all fifty-nine community board districts in the city. Of course, we would lose an economy of scale in managing these small facilities, and some neighborhoods would have trouble finding a place to put them. Still, it may be a good time to develop the technology to make smaller, cost-effective waste facilities. If we all had to manage our own garbage, maybe we'd figure out a way to make less of it.

There are many factors that influence behavior concerning waste production, disposal, and reuse. As we have discussed, energy is one of the major costs of waste management, regardless of the disposal method. With reductions in energy costs, all sorts of garbage mining and recycling might become more cost effective. Another factor that influences the cost of recycling is the price of new materials. The economics of recycling can change when a substance becomes expensive, scarce, or both. Salvage businesses that were unimaginable a generation ago are now profitable. As the management of the waste stream becomes more efficient and cost effective, the cost structure of a city's waste management infrastructure can become more competitive.

The data on waste diversion nationwide indicates that the waste diversion trend line is quite positive. In 1960, 94 percent or 2.51 of the 2.68 pounds of waste we disposed per capita ended up in landfills or non-energy-producing incinerators. In 2008, 54 percent or 2.43 pounds of every 4.5 pounds we disposed of each day suffered a similar fate (U.S. EPA 2008). It is possible to imagine a waste management system where very little of the waste stream ended up in a landfill, and the current trend is certainly in that direction.

For cities, closing the loop on waste disposal is an essential element of a sustainable system of material flow. Of course, each city will have unique elements of its waste management system. Geography in particular will play a key role. A city that is surrounded by suburbs will have one set of options, while one in the desert or on a coast may have another set of options. The local economy and culture will also influence the content of the waste stream and best possible waste management options. It is important for cities to analyze the best waste management practices and technologies for their own circumstances.

Landfills are often the least expensive options, particularly in areas with access to inexpensive land. However, leaking landfills can damage vital supplies of groundwater. While the environmental quality of landfills has improved dramatically, even the "impermeable" linings required by EPA regulations eventually leak (Allen 2001). More importantly, as resource recovery becomes an increasingly common element of waste management, mining the waste stream will become an important element of a city's revenue picture (Douglas 2008). Cities investing in facilities that make productive use of waste instead of dumping it will be better situated to take advantage of that potential source of income. The only mitigating factor is avoiding the adoption of unproven technologies. If you invest in infrastructure based on a new technology too early, you can end up adopting flawed or cost-ineffective technology.

The other issues related to waste are the toxicity of the waste stream and the presence of abandoned toxic waste sites within the city. Older and abandoned industrial facilities (sometimes called "brownfields") can be riddled with toxic waste and be expensive to detoxify enough to be reused. Electronic waste from abandoned computers, TVs, iPods, and other devices can inadvertently transform "normal" solid waste into toxic waste. Here too, cities are dependent on economic trends and regulations beyond their control but that must be accommodated as cities evolve toward sustainability.

The issue of waste toxicity is far from trivial. Toxic substances can be difficult to recycle and can contaminate otherwise nontoxic and valuable substances in the waste stream. Research on detoxification holds potential but has not been emphasized. There are three strategies now in use for dealing with toxic waste. The first is to ensure that people are never in the pathway of exposure. The second is using water and filtration systems to gradually dilute the toxicity of a piece of land. The third strategy is to contain and segregate the contaminated area. Here, you take the toxic substances, remove them from wherever they are, and place them in a "secure" facility, or you build a containment structure around the toxic waste site.

Sustainable Air Quality

The first environmental issue regulated by the federal branch of the U.S. government was air quality. National ambient air quality standards were first mandated by the 1970 version of the Clean Air Act. The air pollution law was followed closely by the Federal Water Pollution Control Act, enacted over President Nixon's veto in 1972. Air pollution is a problem that is easy for the average person to understand. While some air pollution is odorless and colorless, most polluted air is tinted and carries an odor. On the other hand, almost all clean air is invisible and scent free.

While some sources of air pollution are local factories, power plants, and motor vehicles, a great deal of air pollution is carried by air currents from distant communities. Therefore, to some extent, a city must rely on national and even international policy to ensure high-quality air.

Nevertheless, if a visitor or prospective resident of a city gets off an airplane and can *see* the surrounding air, it is likely that she will notice and as a result consider alternative locations for her next visit. Air pollution can also affect the health of local residents, causing illness and increasing the costs of health care.

Local land use policies can improve urban air quality. Mass transit produces less air pollution than individual transit, and green roofs, tree planting, and parkland can all increase a locality's ability to absorb pollution and help reduce pollution generation as well. Nonfossil fuels and energy conservation can reduce the environmental impact of electric power generation, thus diminishing the need for polluting power plants near cities.

The issue of climate change can affect air quality as well. During the summer, heat from vehicles, air conditioning, and other local sources remain unabsorbed by a city's concrete surfaces and can create an urban heat island effect. This in turn can increase air pollution and the level of risk for people with asthma and other respiratory problems. For a city focused on its air pollution issues, it makes sense to work on initiatives that can reduce locally generated impacts (U.S. EPA 2009a).

Quality of Life: Parks, Culture, Entertainment, Education, and Health Care

In addition to the tangible resources described above, cities are ultimately environments that allow people to build communities and friendships, raise families, and live their lives. To sustain a high quality of life, a city must serve as a home to well-functioning cultural and educational institutions and places that amuse and entertain. Health care must be available, as must collective community events that build civic loyalty and pride. Open space, recreation, and other resources that enhance the quality of life are essential elements of the sustainable city. Cities must be places where people enjoy living or want to visit. They must facilitate, not impede, social interaction and the free exchange of ideas. These are not simply frills and add-ons; they are the essential elements of sustainable urban places.

An example of the role of arts and culture in civic life took place in New York City in 2005, when the New York artists Christo and Jeanne-Claude set up their "Gates" exhibit in Central Park. According to their Web site:

> The installation, at the site in Central Park, was completed with the blooming of the 7,503 fabric panels on February 12, 2005. The 7,503 gates, 16 feet (4,87 meters) tall varied in width from 5 feet 6 inches to 18 feet (1,68 to 5,48 meters) according to the 25 different widths of walkways, on 23 miles (37 kilometers) of walkways in Central Park. Free-hanging saffron colored fabric panels, sus-

> pended from the horizontal top part of the gates, came down to approximately 7 feet (2,13 meters) above the ground. The gates were spaced at 12 foot (3,65 meter) intervals, except where low branches extended above the walkways. The gates and the fabric panels were seen from far away through the leafless branches of the trees. The work of art remained for 16 days, then the gates were removed and the materials industrially recycled.
>
> (Christo and Jeanne-Claude 2005)

Christo and Jeanne-Claude first proposed this exhibit in 1979 but were turned down by the city, which was still recovering from its near bankruptcy five years earlier. This time around, the arts booster and mayor Mike Bloomberg agreed to allow the exhibit. The artists would have to raise the funds to pay the costs, but New York City would cooperate. According to the New York City Mayor's office:

> The Central Park Conservancy's attendance count estimates that visits to Central Park reached over 4 million during *The Gates*—a substantial increase from the approximately 750,000 visits the Park receives during the same two week period in a typical February. NYC Economic Development Corporation estimates that more than 1.5 million visitors for *The Gates* were from out of town—an estimated 300,000 of those were international visitors. Usually 13% of tourists are from outside the country but during *The Gates*, the international share increased to almost 20%.
>
> (Office of the Mayor, City of New York 2005)

Perhaps the impact of this visionary and whimsical exhibit could not have been predicted, but it captured people's imagination and was a social sensation; New York City's government estimated the economic impact of the Gates at over $250 million (Office of the Mayor, City of New York 2005). These types of initiatives make a place unique and distinctive and add to a city's quality of life.

Educational and cultural institutions employ and attract people to the city for their programs and offerings. Educational institutions are particularly crucial for training workers and developing intellectual properties that can create new expertise in a high-technology economy. Michael Crow, the president of Arizona State University, has explicitly connected university research and development expenditures to statewide economic development in Arizona (Crow 2010). Moreover, the university, which has its main campus in Tempe, a suburb of Phoenix, has developed a campus in downtown Phoenix. The goal of the new campus is to help return the residential community, nightlife, and urban amenities that disappeared from Phoenix a generation ago back to its downtown (City of Phoenix 2010). Put thirty thousand students in a neighborhood, and bars, coffee shops, and bookstores will sprout like weeds.

As the technology of health care advances, more and more of our GDP is devoted to health care, and our expectations of access to quality health care grows. For a city to attract and retain residents, it must have a well-functioning health care system (Adams 2005). A city's health care institutions are as critical an element of its infrastructure as its roads and water pipes.

The fundamental questions for urban sustainability are: Why should someone move here or stay here? What does this city offer? Some of the resources that cities require deal with basic biological necessities such as food, water, air, and waste. Others have to do with the quality of life after those necessities have been met: culture, education, entertainment, and so on. From the museums to the theater district to the bike path along the river, sustainable cities must offer a little bit of everything. Ours is a mobile society, and people who do not find one city's quality of life adequate are more than capable of picking up and moving elsewhere.

Financial Issues

As America struggles to make its cities more energy efficient and build mass transit, electric car charging stations, smart grids, sources of renewable energy, modern water systems, and even schools and roads, we face financial challenges in every way. Travel through Europe or Asia and you will see high-tech airports and state-of-the-art bullet trains; come to New York City and you see the rundown JFK Airport and the converted warehouse we call Penn Station. While the infrastructure finance crisis in 2010 could be blamed on the current recession, the longer-term problem is that Americans are undertaxed and that we do not save enough of what we earn. I admit that I dislike paying taxes as much as the next guy—and I spend lots more than I save. But the reality is that there is insufficient money for infrastructure because we are unwilling to invest in the future.

One of the central issues for sustainable cities is who will pay for the transition to a green urban economy. In particular, how is public infrastructure to be financed? To some degree, this should be a mere continuation of the role that local governments in the United States have long played in financing infrastructure projects. Taxes and user fees generated at the local level can be used to create the revenue streams needed to pay off debt from capital expenditures. More creative approaches to financing have been developed in recent years. In both California and New York, surcharges on electric utility bills fund energy efficiency and renewable energy expenses. In a number of American states, Property Assessed Clean Energy (PACE) bonds have been developed to fund energy efficiency and renewable energy projects. PACE programs enable property owners to pay for these energy improvements over fifteen to twenty years via an increase in their annual property taxes (PaceFinancing.org 2010).

Where sustainability investments can save money, creative forms of municipal finance can transform future savings into capital for immediate investments. In many cases, this can work, but in others, we need to be creative, as new services and revenue streams will be needed. In recent years, we have seen this take place, in some respects. At one point, the cost of water delivery was hidden in property taxes, but in recent decades it became a separate charge in many localities. At one point, entertainment was freely available to anyone who owned a TV or radio and an antenna. Today, we pay monthly bills for cable television and satellite radio. Urban amenities that promote sustainability will also carry charges in the future. Congestion pricing for driving in a crowded central business district is one example; I suspect we will invent many others.

The total tax rate in the United States is much lower than that of most European nations. In 2006, total taxation in the United States came to 28 percent of our GDP. This is compared to 44.2 percent in France, 37.1 percent in England, 35.6 percent in Germany, and 33.3 percent in Canada. Only the Japanese and Koreans pay slightly less taxes than we do (Bartlett 2009). In the United States, we face the challenge of financing the infrastructure needed for the twenty-first century while paying for the military required for our national defense. Despite assertions to the contrary, our problem is not excessive military spending but that we are undertaxed. The United States spends about 4 percent of our GDP on the military, compared to 2.6 percent in France, 2.4 percent in England, 1.5 percent in Germany, and 1.1 percent in Canada. Even if the military spending of these other nations equaled ours, we would still have less funding than they do (as a percentage of our GDP) for domestic public services (Central Intelligence Agency 2010).

In addition to being undertaxed, until the Great Recession of 2008–2010, Americans were consuming more and saving less. The percentage of disposable income saved by Americans dropped steadily from about 10 percent between the mid-1960s and the 1980s to less than 4 percent in the 1990s. At several times in the early twenty-first century, the American savings rate fell to zero and below (Rampell 2009). Coupled with low taxes, we saw a financial picture of a society bingeing at the mall instead of saving and investing in the future.

It is not that increased private saving directly translates to more funding for public infrastructure but rather that it indicates that our culture may be more receptive to long-term investment. During the credit-driven boom years of the early twenty-first century, consumption was both king and queen. People leveraged their homes and retirement savings to buy massive SUVs, big-screen TVs, snazzy vacations, and countless other products. Meanwhile, the capital needs for basic research and development and national infrastructure were ignored. The Great Recession and the stimulus packages that followed resulted in increased rates of individual saving and a growth in public investment in green infrastructure. For many people, the shock of losing half of their retirement funds resulted in increased thrift and savings. As the high-end retailers in 2009 and 2010 on New York's Madison Avenue or on Los Angeles's Rodeo Drive could tell you, the party is over.

However, because tax rates did not increase in 2008 through 2010 (to avoid turning the recession into a depression), infrastructure investments made with U.S. federal stimulus funding were not built on a sustainable financial footing and have only increased the national debt. Cities and states in the United States face multiple fiscal crises, and despite federal funds, investment in sustainability initiatives has stalled in many places.

In the long run, investment in the transformation of our cities into environmentally sustainable communities will require capital from a variety of sources. One source will be debt retired by increased taxes. Another source will be savings generated from more energy and resource-efficient technologies and management practices. Some of this will be market driven, as more resource-efficient companies gain the ability to offer lower prices than those that are less capable of controlling their use of resources. However, some of the investment in sustainable cities will need to come from public sources or at least be driven by public incentives. Financial issues must be addressed if sustainable cities are to become a reality. If we keep starving infrastructure investment, we will fall farther behind the global competition. Until these issues are addressed, the idea of a sustainable city will be more of an aspiration than a living, breathing place.

Organization and Management Challenges

National governments make policy and collect money, but local governments are where the "rubber meets the road." City governments vary widely in capacity, structure and competence, but all city governments deliver essential and nonoptional services like supplying water, collecting garbage, and assuring public safety. Many cities and towns in the United States have developed sustainability plans, and in many communities there is enormous pressure from the grassroots to reduce waste and develop more environmentally conscious practices in waste, energy, water, transportation, and land use. Support for sustainability policies is strong among young people and particularly so on college campuses. Cities have also begun to move in this direction. Through zoning, cities have long influenced land use policy, and larger cities have worked to shape local economic development. These are not new functions for local governments.

In the case of sustainability management, additional planning capacity and an enhanced ability to work with private businesses are capacities that are sometimes lacking. Within many private organizations, expertise in energy reduction and resource use must be upgraded. Organizations must combat the tendency of certain users to overbuild in order to avoid failure. A prime example can typically be found in an organization's information technology unit. Data farms and servers are notorious energy hogs. For information technology managers, no price is too high to prevent failure, so they build multiple redundancies into operations. Some of these make sense, but others are excessive to the point of true absurdity (O'Carroll 2005). Similar tendencies are found in heating and air conditioning systems and in other parts of facilities management. Senior management causes this behavior through their actions in response to system failure. While I am not arguing for more fragile essential systems, a number of firms have developed expertise in cutting resource costs without significantly raising the risks of failure.

Organizational inertia in the public and private sector must be overcome if we are to develop the capacity to manage our cities sustainably. Sustainability plans and other policies are necessary but insufficient conditions for the development of more sustainable urban places. Just as it took many years for modern accounting and financial control systems to replace less rigorous and informal methods, this type of organizational change will take many years to implement.

Public and private organizations must do a better job of communicating intent and practices and coordinate their actions with one another. This will require an environment of trust and accommodation. A good example of public and private cooperation is the conversion of driving streets into pedestrian plazas. When a number of streets became green spaces, local businesses first complained about the inconvenience and potential effect on sales. After about six months, they found that slower-moving and less-stressed pedestrian traffic actually helped increase business. Innovative changes can be hindered by the mindset of "if it ain't broke, don't fix it." For change to occur, stakeholders must believe that they can influence the creation and outcomes of a new program. This requires that resources be devoted to sophisticated interaction and communication with stakeholders.

Government Policy

The market alone will not produce sustainable cities. Government alone cannot produce sustainable cities. It will take partnership, collaboration, experimentation, investment, and trust. We need to identify and encourage the organizational and individual behaviors that will reduce waste and consumption while maintaining economic growth and quality of life. While government can encourage, set boundaries, and educate, real change is nearly always a reflection of entrepreneurial self-interest. A system built on the profit motive makes things happen. Sustainability management must make money for the people who adhere to its principles. It must allow sustainable parties to make more money than parties who do not pay attention to sustainability. Government must therefore set rules of the game that encourage sustainability practices.

Just as speed limits, seatbelt laws, and air bags are designed to regulate traffic safety, sustainability practices must be defined and encouraged through the rule of law. Government is not suited to make autos, but it can regulate their safety levels. Government policy develops neighborhood- and citywide rules and infrastructure to encourage, facilitate, and occasionally mandate sustainable practices.

In many cities, government units that work to encourage economic development have added sustainability planning units. Planning boards and zoning commissions, transportation, sanitation, and environmental agencies are also adding sustainability-oriented units to their organizations. Cities are growing all over the world, and as they grow they have the opportunity to steer new development and reconstruct old structures to be more material, energy, and water efficient. They can also work to reduce effluent and emissions discharges as they develop "closed-system" facilities. Government policy can encourage these behaviors with tax incentives and discounts on water and energy charges. While this may seem unrealistic in the impoverished shantytowns of the developing world, much of the development now underway is not in these horrific slums, and when those slums are replaced by decent housing, there is no reason why new development cannot adhere to the precepts of sustainable architecture.

The other crucial government function for sustainable cities is to develop citywide plans to guide development and invest in or encourage private investment in transportation, energy waste, and water infrastructure. This is not an easy task, and as Scott Campbell points out in his 1996 article for the *Journal of the American Planning Association*, "the planner must reconcile not two, but at least three conflicting interests: to 'grow' the economy, distribute this growth fairly, and in the process not degrade the ecosystem" (Campbell 1996, 3). Government's engagement must be strategic and thoughtful, always keeping in mind the limits of government's role and the need to draw upon community-based and private market forces to take action.

Chapter 7

A Sustainable Planet

Definition and Technical Challenges

Sustainability is the word of the moment and certainly a term explored in some detail in the preceding pages. We began by describing the meaning of sustainability in business and moved on to a variety of resources, including food, water, and energy. Then we integrated the discussion to consider the place where most people now live: the planet's cities. This chapter will discuss the issues of global scale and a sustainable planet. I will raise many of the same sustainability issues raised in earlier chapters, but with a different focus or unit of analysis. This chapter looks at the degree to which individual sustainability issues either on their own or in combination can have effects on a global scale. For example, the destruction of an ecosystem can impair the quality of water in a local watershed, but the impact on the watershed is not a global issue. On the other hand, the destruction of a species in the ecosystem could have a long-term global impact that we do not yet understand.

In some respects, all local issues have the potential of scaling up to the global level, and all global issues begin with place-based impacts. When we discuss sustainability issues that have a widespread impact on global systems, it is best to imagine a continuum. On one end of the continuum we have impacts that are purely local, in the middle we have those that can affect other localities, and on the far end we have impacts like climate change, which affect planetwide functions.

I should also mention that planetary sustainability does not require that we set the bar so low that we are only concerned with the planet's survival. Our criteria and definition of planetary sustainability should enable us to obtain an *acceptable* quality of life level for all of the planet's inhabitants—particularly the human ones.

In this context, we are trying to understand how to manage and steer the impact of people and their productive capacities on the planet's natural systems. There are a variety of reasons why we should be concerned with the viability of the planet's natural systems. Most fundamentally, as biological creatures, we need the food, water, and air of the planet in order to survive. We need to ensure that the quality of these resources will allow us to not only survive but to thrive. Someday, we may have the technology to create our own resources with the application of technology and limitless energy, but we do not have the technology or energy sources to do so at this time.

However, let's say that some day we do have the replicators and the "warp" reactors we see on *Star Trek*. We would not need the planet to physically survive, but would the freedom to destroy this place entitle us to do so? At that point, sustaining the planet is no longer a common-sense response to our desire to survive but a matter of choice and values. At that point, sustainability becomes an issue of preserving the planet's great beauty and an expression of our respect for the miracle of its creation. I doubt anyone alive will live to see the moment when humans no longer need the planet but instead preserve it out of love, respect, and even veneration, though it is easy to imagine reaching such a point at some time in the evolution of our species.

Science fiction provides us with images of such a future, and if history is a guide, if we can imagine it, some day we will probably be able to build it. From da Vinci's drawings of flying machines, to horseless carriages, to the flip phones used in the early *Star Trek* episodes, human technology has a way of following human dreams and aspirations. Unfortunately, as the atom bomb taught us, human nightmares have a way of becoming reality as well. When thinking about the preservation of our planet, consider that the home planet in the *Star Wars* films is a world city of many layers of development—where nature has been completely banished.

Fortunately—or unfortunately—such technological marvels are beyond our capacity. It has been over two generations since humans walked on the moon, and for the time being we must preserve this planet and its resources or suffer the consequences. At the same time, I hope that an appreciation of the intrinsic beauty and wonder of this planet is established so that when we get to the point when the planet is no longer needed for our survival, our appreciation of it will cause us to preserve it for its own sake.

A general, overarching challenge of achieving a sustainable planet is to develop a sufficient understanding of Earth's systems to manage them. This includes the creation of an Earth observation system that measures natural changes in the planet, human-influenced changes, and the effectiveness of measures designed to mitigate damage. The controversy over climate science is an indication of how crucial observational earth science and defensible models are to sustainability policy and management. Such an observational system is not yet in place, although pieces of it have been developed. Remote sensing, satellite images, and advanced communication and computational technologies hold the promise of reducing the costs of effective Earth observation.

In addition to the issue of sufficient Earth observation, we also need to develop technologies that enable us to preserve habitats and biodiversity; control, store, and possibly recycle global warming gases; treat wastewater to enable reuse; and detect the presence of weapons of mass destruction. Finally, we need the organizational capacity to manage sustainability on a global scale. We need organizations capable of conducting Earth observations and developing and building the technology and practices needed to mitigate human impacts on the biosphere. Specifically, we need to learn how to protect and preserve biodiversity and ecosystems, prevent and adapt to climate change, protect and improve air and water quality, ensure a sufficient food supply, develop land and dispose of wastes with the least possible impact, and prevent wars, terrorism, and dysfunctional human conflict.

Protect and Preserve Biodiversity and Ecosystems

The complex and interconnected web of our ecosystems is the fundamental source of human wealth and well-being. Some of the maintenance of these systems simply requires that we leave them alone. But in other instances where human-induced impact has already taken place, we need to learn how to restore existing ecosystems.

All over the world, population growth and development is putting the biosphere under constant stress. In what I consider a classic article in the February 24, 2000, issue of *Nature*, entitled "Biodiversity: Extinction by Numbers," Stuart L. Pimm and Peter Raven attempted to estimate the pace of species extinction now under way. Pimm and Raven (2000, 403) observe:

> Humanity is rapidly destroying habitats that are most species-rich. About two-thirds of all species occur in the tropics, largely in the tropical humid forests. These forests originally covered between 14 million and 18 million square kilometres, depending on the exact definition, and about half of the original area remains. Much of the rain-forest reduction is recent, and clearing now eliminates about 1 million square kilometres every 5 to 10 years. Burning and selective logging severely damages several times the area that is cleared.

The number of species and the diversity of genetic material within these species depend on ecosystem health for their own survival. We don't actually know how many species there are on this planet—estimates range from two to ten million—but we know that we have not yet identified all of them. Our knowledge of the ecosystems that support this biodiversity is also very incomplete. We are destroying ocean, forest, and freshwater habitats worldwide at a rapid pace. The economic pressure to develop land can only be resisted when it is taken out of circulation by private purchase of land for purposes of conservation or public park use. Nonprofits such as Conservation International have purchased huge amounts of land for preservation, and organizations such as the Nature Conservancy work with governments to preserve land. In some cases, Conservation International or governments do not purchase land but instead purchase the land's development rights and allow owners to use the land without developing it.

These strategies have been effective in preserving ecosystems on land, but oceans are a different matter. Most of the planet's surface is covered by oceans, and since they are generally not within the jurisdiction of nation-states, like the air, they are a form of global common property. The oceans are under enormous and relentless stress partially because of their function as a commons. The benefits of fishing are direct for the individual, while the cost of overfishing is indirect and paid by all of us. Fisheries have been decimated, and many aquatic environments have been damaged or destroyed. Marine debris is increasing in weight and toxicity, and invasive species are carried in the ballasts of ships and dumped in foreign waters. No one has taken ownership of these problems. The issue of global governance comes into play when we think about sustainable oceans. Although we have an international law of the sea, it is clearly not effective in preventing environmental damage. Without binding and enforceable environmental ocean rules, the oceans will continue to deteriorate.

The problem of ecosystem and biodiversity preservation is all around us. Some variants are quite visible, but other issues are in distant locations or close by but less visible. For the planet to be truly sustainable, we will need to increase our understanding of ecosystems and decrease our negative impact on those systems. We will need to build the organizational capacity to implement mitigation strategies.

Prevent and Adapt to Climate Change

Experts on paleoclimate understand that the planet's temperature is constantly changing and has done so throughout history. Today, the scale of human impact on the planet is so great that in addition to these natural climate cycles we have begun to see human-induced climate change. These human-created impacts have increased the pace of climate change and have already made the planet significantly warmer. None of this would matter a great deal if we did not have a great many people living on the planet as well as constraints on immigration and trillions of dollars invested in infrastructure that assumes specific conditions of the land, air, and water. Sea-level rise due to glacial melt could damage coastal cities. Changes in weather and precipitation patterns can reduce the productivity of cropland and disrupt communities and nations.

There are so many greenhouse gases in the atmosphere that climate change has already begun and will become increasingly intense throughout the twenty-first century. As my colleagues Klaus Lackner and Wally Broecker have written, we will need to sequester and store greenhouse gases, attempt to reduce the emission of new gases, and learn to adapt to the changes we have already set in motion. The climate issue has become a sort of symbolic dividing line between environmentalists and skeptics, and the emotional content of this issue seems to have grown. Part of the problem is that addressing this issue is difficult and will require a long lead time. Often advocates predict future impact with greater certainty than any probability statement should allow. Another aspect of the climate issue is that the worst aspects of the problem lie in the future, and unlike other environmental issues, it is not possible to see, smell, or feel climate pollution. Toxic waste oozes, air pollution smells, and many water pollutants can be seen and smelled. Despite the views of the "climate change deniers," the science of climate change is unambiguous. While the planet can survive as it gets warmer, the adverse impact on human settlements, ecosystems, and accumulated infrastructure and wealth will be dramatic.

At a global level, the issue of climate change appears to pose profound challenges to our international political regime. Climate change is one of the most vexing examples of the tragedy of the commons. This is exacerbated by the long-term temporal dimension of the commons that we are overusing. In the classic example of the overgrazed common field, you can see the green of the field turn to brown in a matter of months or growing seasons. In the case of climate, the most severe impacts may be decades in the future and may be most profound in a distant location. National governments with sovereignty over land have difficulty with long-term, amorphous issues such as climate change. International organizations with less power and resources are even more helpless to directly address the need to reduce greenhouse gas emissions.

However, unlike ecology and biodiversity, the issue of climate change is actually easier to understand and more amenable to a technological fix. Ecosystems are complex, intricate, and difficult to measure. As indicated earlier, we are still discovering new species, and many other aspects of ecosystems remain unstudied. The climate issue is actually more straightforward, and it will be solved when we perfect the technology of renewable or nuclear energy and its lower price eliminates the widespread burning of fossil fuels. The low-carbon green economy is within our reach technologically. As pressing as the climate issue may seem today, the longer-term challenge of preserving species diversity and functioning ecosystems will probably face us for a longer period of time.

Protect and Improve Air and Water Quality

Returning to the theme of the biological nature of all living beings, human biology requires clean and sufficient air and water. Modern production processes and human settlements can be managed to the degree that impacts on water and air quality become minimal. The past half century or so of intense population and economic growth have been matched by dramatic advances in the technology of industrial production, water filtration, air pollution control, sewage and waste treatment, and methods of recycling and reusing raw materials. In the United States, Japan, and Europe, the historical link between economic (GDP) and pollution growth was decoupled in the 1980s. Moreover, in many areas, pollution loads have been significantly reduced.

Urbanization and population increases stress air and water resources but are amenable to technological fixes. Interestingly, one of the major costs of pollution control is the cost of energy. If energy costs were to be reduced, the costs of filtering air and water for reuse would also come down. The geography of water and air pollution includes cross-jurisdiction and within-jurisdiction dimensions. For that reason, effective local and national rules and management are required along with international rules and enforcement.

The concept of airsheds and watersheds tells us that to some degree the problems of clean water and clean air are local rather than global problems. However, these are resources that are influenced by environmental changes at the global level. Climate change affects the weather and water supply and can exacerbate air pollution problems. Both clean air and clean water depend on the filtration and environmental services performed by well-functioning ecosystems. In many ways the modern environmental movement began when Rachel Carson's landmark work, *Silent Spring*, alerted the world about the inadvertent damage caused by DDT to biological entities it was not targeting. Shortly after her work, Barry Commoner's *The Closing Circle* reported that radioactive fallout from weapons tests in the Pacific Ocean could be found in the milk of cows thousands of miles away. While the Earth is a pretty large place, air pollution must be seen as a global problem rather than a purely local one.

Air and water pollution require that we think globally and act locally to echo the sentiment once expressed by Rene Dubos. When thinking about a sustainable planet, we need to think about global-level environmental problems that are caused by local issues throughout the world. Air and water pollution are examples of that sort of problem.

Ensure Sufficient Food Supply

Food was once largely locally grown and consumed. But starting with coffee, tea, and some alcoholic beverages, in the seventeenth century we began to see the start of a global food trade. Today the food industry is a huge, multibillion-dollar multinational business. As the planet urbanizes and people are no longer involved in food production, the food supply system is largely invisible to the average citizen. The grapes you eat could come from one hundred or one thousand miles away. The food supply system, like many of our systems of material consumption and production, is increasingly global. Multinational food companies produce and distribute the food, and food consumption habits have come to transcend growing seasons and geography.

A sustainable planet requires that food be produced in a way that enables it to continue year after year. Sustainability also requires that food supplies increase to keep pace with population growth. It was Malthus's original computation about the balance of food supply with population that led him to predict that food production would fall behind and starvation would result. Had technology not developed new methods of food production, it is likely that his prediction would have come true. A third issue is that food distribution must not be completely concentrated in wealthy nations. Starvation and malnutrition in the developing world and obesity among the poor in the developed world create serious public health problems and can even destabilize a nation's political institutions. Poverty and particularly the inability to feed one's family can lead to political desperation, instability, and even terrorism. This is not to excuse those responses, but wealthy nations that want to preserve their wealth need to put a floor under human poverty and prevent starvation.

As the chapter on sustainable food indicated, the production of food involves energy, water, fertilizer, pesticides, seeds, feedlots, and waste management. It is a resource-intensive process, and its sustainability requires the ready availability of those resources. Scarcity in parts of the food system can increase prices and reduce access. Because all food comes from living entities—plants and animals—the ecosystems upon which they depend must be preserved to prevent us from starving.

Develop Land and Dispose of Wastes with the Least Possible Impact

Food is part of a global system, and while waste tends not to be distributed on a global basis, some of it is shipped far from home, and some "recycled" products are shipped thousands of miles away to be dismantled and scavenged for component parts. Poor people can find themselves exposed to toxic chemicals when they dismantle computers or other electronic products for parts.

The idea of taking solid and toxic waste and dumping it into a hole in the ground is a traditional method of waste management. If humans survive long enough on the planet and our population continues to grow, eventually we will be living in our garbage. Simply keeping track of all this waste could overwhelm our capacity. The very first major toxic waste emergency was the environmental disaster at Love Canal, in Niagara Falls, New York. The release of toxins into a schoolyard and nearby homes was a result of the burial of barrels of toxic waste in an abandoned canal. After a cold winter and a wetter-than-usual spring, the barrels became unearthed and began to leak. As we do not have the technology to detoxify land at the present time, this type of waste disposal will have the affect of reducing the amount of land we can use for more productive purposes.

In addition to using it to dump garbage on, we use land to live on and to grow food. Humans live in a wide variety of settlements, from cities like New York and Hong Kong to small villages in Africa. Some people live in huts, and others live in mansions occupying acres of land and structures running into many thousands of square feet. The constraints on decent housing are not the planet's land mass or the availability of raw materials to build dwellings. There are largely political decisions related to the distribution of wealth.

While the quality of housing and the toxicity of the land are key issues, another issue that affects global sustainability is the impact of human settlements on ecological systems. Careful design and an understanding of the composition of surrounding natural systems can dramatically reduce the environmental damage caused by our cities, towns, and farms. While land use and waste issues are local and not global, given the billions of people on the planet and their growing patterns of material consumption, the gradual impact of destructive land use practices and poorly thought through waste management practices can have a global effect on these ecological systems.

A good example of the global impact of local pollution is the spread of disease and the problem of invasive species. Due to worldwide commerce, shipping, and travel, diseases created in one part of the world are easily transported in human, plant, and animal hosts to other parts of the world. Some of the natural predators of some species and some of the evolutionary resistance to particular diseases are absent in new settings, and the effects on human and ecosystem health can be dramatic. Some anthropologists believe that one of the main causes of death of the Native Americans during the European settlement of America was the intensity of the diseases brought by settlers to the New World from Europe. Diseases such as the swine flu and SARS are examples of global diseases that have substantially challenged modern medical technologies and public health management practices.

Another issue related to land use is that the planet's growing population and areas of greater density increase the probability of destruction of human settlements from natural disasters such as earthquakes, hurricanes, and tsunamis. If one compares the earthquakes in Haiti in late 2009 to the one in Chile in early 2010, we see that the destructiveness of the quake is a function of its intensity, location relative to cities, the rigor and enforcement of local building codes, and the capacity for effective emergency response. As we saw during Hurricane Katrina, emergency response capacity is a necessary but not sufficient condition for effective emergency response. There must also be the political will to rapidly deploy this capacity if the negative impact on lives and property is to be as low as possible.

One of the problems with patterns of land use is that once they have been established, barring natural or human-made disasters, they are very difficult to reverse. Once transportation, energy, and water infrastructures are installed, land use development tends to follow the path of that installation. In major cities with subway systems, you see higher population densities and much more commercial activity by train stations. In places with superhighways, you see greater development presence (albeit more spread out in form) by highway exits. These patterns can be changed through infrastructure investment, and the quality of drainage and filtration systems as well as the location of built and natural environments can have a major influence on environmental impacts.

Prevent Wars, Terrorism, and Dysfunctional Human Conflict

When thinking about the massive destruction of ecosystems, we sometimes think about the gradual destruction of these systems as we produce the food and material goods that sustain us and the wastes that result from that production. In that respect, we can justify the destruction as a trade-off that we've made in order to produce the things we require for survival. But there is another more rapid and devastating form of destruction, which is the type that results from human conflict. Wars, terrorism, and other forms of violence target people but may also target natural systems. The idea is to destroy an enemy's capacity to make war. When Sherman marched through the South near the end of the Civil War, his troops burned farms, structures, and anything with productive capacity.

Modern weapons of mass destruction hold out the potential to destroy entire parts of the globe and make them uninhabitable for centuries. Self-interest and competition are seen by some as inherent human behaviors, hard wired into our genetic material as a result of Darwinian selection resulting from ancient struggles for survival. The idea is that in the Hobbesian "state of nature" of the war of all against all, the strong survived and the weak perished, and so now those traits dominate the gene pool. Of course, others argue that along with this aggression, intelligence was also naturally selected for—along with the development of rational self-interest. This in turn led to alliances and cooperation. Whatever our inherent human makeup, we now possess weapons that could destroy humanity if they are ever fully deployed. If we end up destroying human life on this planet, we will probably also manage to destroy much of the rest of the biosphere as well.

This is why we must learn to contain, control, and resolve human conflicts without resorting to the use of weapons of mass destruction. There is some evidence that the taboo on the use of nuclear weapons, set after the unforgettable fires of Hiroshima and Nagasaki, is firmly in place. These weapons have not been used since the end of World War II. Unfortunately, their use is not unimaginable. The nightmare vision is the possibility of a terrorist group developing the means to detonate an atomic bomb in a major city. While a single explosion would not destroy the entire biosphere, the resulting political instability could very well result in multiple detonations with unpredictable impacts.

Preventing the use of weapons of mass destruction requires resources, organizational capacity, and technological innovation. The "Star Wars" antimissile defense system advocated by the Reagan administration in the 1980s was an effort to develop a technological fix to one of the problems caused by weapons of mass destruction. Of course, as soon as one weapon is countered by another, it is not long before another technology is developed to counter the new technology. The endless arms race should focus us on the need to address the fundamentals of conflict and move the resolution of disputes into nonviolent arenas.

In the final analysis, all the progress we might make in the other arenas of sustainability could be undermined by a war fought with weapons of mass destruction. This means that this issue of global security dominates all other issues and is the single most important of all sustainability issues.

Financing a Sustainable Planet

Much of the resources needed to make our planet more sustainable will simply require that we direct money we are now spending on management practices that are not sustainable to those that are. Private market forces have already begun these processes. However, some funded needs will not be substitutional but incremental. In particular, funds must be found for the transition. Government will need to generate and then allocate resources needed for sustainability infrastructure. Some infrastructure capital can be raised by utilities that are given monopolies in return for the obligation to provide collective goods for a fee. I can imagine that a smart grid might be funded by electric utilities—ultimately with savings generated by the more efficient distribution of electricity.

More efficient use of resources and the use of renewable, solar-based technologies hold out the promise of reducing the percentage of the gross domestic product devoted to energy generation and distribution. In most cases, a more sustainable use of a natural resource will reduce the long-term costs of using that resource. The problem is the notion of "long-term" savings. All of our metrics in management, politics, governance, and even sports and entertainment are reported annually or even more frequently. Long-term investments are difficult to generate or justify in the current process. Managers and leaders understand that they should take the long view, but current incentive structures do not reward long-term thinking, and thus few managers indulge in it. Occasionally, you will see a patient and principled investor such as Warren Buffet, who takes a long-term perspective and resists the herd. But they are rare and far from the norm.

Much of the public-funding need is for collective infrastructure such as water, waste, transport, and energy systems. The U.S. public's antipathy to anything that is remotely labeled as government taxation will constrain the organizational and ownership form of these collective resources. While the public complains about and resists higher water fees and the fuel taxes that might fund personal and mass transit infrastructure, at the same time they are voting with their checkbooks by spending more and more on their cable, Internet, and energy bills. To the degree possible, we will see the U.S. federal, state, and local governments delivering infrastructure as part of regulatory and privatization deals with private firms.

If properly governed and policed, this can work. Access for people with lower income can be required, and other public policy goals can be achieved in exchange for monopoly access or limited competition and in some cases guaranteed rates of return on capital. In most respects, these arrangements work best when private-sector competition is real and vibrant. Fat and bureaucratic cable TV outlets have been forced to improve their prices and service when challenged by satellite TV competitors. Cell phone companies have had to respond to the challenges of the Internet. However, corruption and sweetheart deals are also a side effect of the privatization game, and so transparency and vigilance are required for this to work. That is true when the capital is raised and spent within the public sector as well. Centralized government control of infrastructure finance and construction in China is far from free of corruption.

While it is difficult for people in consumer-oriented societies to save enough capital to pay for long-term investments in sustainability infrastructure, privatization can result in some diversion of capital for long-term needs. Instead of putting their money in the bank, consumers pay higher fees for energy, water delivery, and waste removal. The companies or, in some states, quasi-government authorities that perform these services use some of the fees to pay for the cost of capital for modern, sustainable infrastructure. This has some of the same capital-formation effect of personal savings and can even be seen as a form of forced savings or even not-so-hidden taxation. It differs from pure taxation because it is a fee for a service, and individuals are free to consume less or "operate off the grid" if they can.

While these techniques can work in the developed world, they are of course less useful in the developing world, where there is not enough surplus wealth to fund either capital formation or increased private consumption. The issue here is whether developing nations can skip the dirty, wasteful, and costly stage of development that the United States saw in the nineteenth and twentieth centuries. When the Western world developed, there was no previously developed world that sat by our side. While initially there was a gap between the United States and Europe in many respects, that gap was quickly eliminated due to the wealth derived from America's rich natural resources. Today, the developed world has the wealth and technology to help the developing world avoid some of the mistakes of our past. While there are some symbolic gestures in the direction of providing such assistance (such as the United Nation's Millennium Development Goals), there is very little evidence that the developed world is currently planning to do much to finance sustainability.

Organization and Management Challenges

Money is a necessary but not sufficient condition to achieve sustainability. In addition, we need organizational capacity. The way to think of it is to bring into all of our organizations, be they government, private corporations, or nongovernmental/nonprofit organizations, the capacity to analyze environmental impacts and natural resource use and act to reduce their impacts and the generation of wastes. We will know this is happening when sustainability analysis and management is considered a management tool akin to accounting systems, information management, and human resources. Modern accounting standards did not develop overnight in complex organizations. These standards evolved over time and are still evolving. During New York City's financial crisis in the mid-1970s, the city lacked an accounting and financial management system accurate enough to tell them how much money they had or precisely what they were obligated to spend. Many types of performance data that we now take for granted were not collected or analyzed in the days before inexpensive computing, the Internet, and cellular communications. However, even the most tradition-bound organizations have proven to be capable of adapting and changing as the definition of effective management has evolved.

I believe that over time these sustainability management capacities will be developed and that they are actually being developed today. Well-managed organizations have begun to understand the need to develop these capacities, even if they do not quite know what they need and how to get there. In the second decade of the twenty-first century, we see increased recognition of the importance of energy and water efficiency and waste reduction. Other related practices such as pollution prevention and closed-system industrial ecology analyses are also being developed and utilized. Sustainability management is in its infancy and will be driven by concerns about costs, availability of resources, concerns about liability, and even our emotional attachment to this planet. As in any new field, those of us looking at it today will not be able to predict its pace, shape, or trajectory. But the factors leading to its development are in place.

This work will take place within local public, private, and nonprofit organizations but will also operate within interorganizational networks operating at a global level. The public and nonprofit sector has an enormous amount to learn from private organizations that have developed mechanisms for operating internationally. The use of local partners and adapting business models and practices to local considerations is commonplace in successful private organizations. Governments and NGOs can learn from these experiences and from the emerging sustainability practices adopted globally by Wal-Mart, Pepsi, Coca-Cola, and a growing group of private firms.

My best guess is that we will see changes in accounting systems to more accurately monetize the benefits and costs of sustainability practices. Tax laws and other regulatory structures including land use rules and zoning will come to reflect these new considerations. Building code requirements such as the energy audits now required in all New York City buildings over fifty thousand square feet will become more common. Green finance tools such as the PACE program (see page 52) will also be expanded.

We will also need to develop a new kind of management capacity in modern organizations, which will be led by people trained in the field of sustainability management. Most people who lead large and complex organizations are either lawyers or MBAs. The lawyers are not trained to manage organizations, and the MBAs learn management and finance but have no training in the physical dimension of sustainability. They do not know the science of environmental impacts or the engineering of energy or resource efficiency. Sustainability managers must know enough of each of these fields to draw on and translate the work of experts. They also must be able to build staff capacity to analyze and implement organizational changes designed to reduce resource use while building production.

The organizational capacity for managing sustainability requires an ability to understand and analyze new technology, analyze and apply public policy rules and incentives, analyze financial costs and benefits, and manage organizational change. This requires constant learning, analysis, and teaching. It requires an understanding of a variety of fields and appreciating how to draw in and elicit work from many types of experts. At the global level, it also requires the ability to interact with partners from all over the world and benchmark operational lessons from distant locations.

The growth of the global corporation and the moves in the past decade by universities and NGOs to expand internationally has created some of the organizational capacity needed for global sustainability management. In 2010, there were four billion cell phones in the world. People without indoor plumbing and electricity in their homes own cell phones. Consumer products are designed in one place, made in several places, and assembled in other places. Our ability to integrate production processes across borders is a main new fact of the late twentieth and early twenty-first centuries. This demonstrates the capacity for technology to penetrate all corners of the planet and is an example of the ability of human organizations to implement actions on a global scale. The same capacities that are used to design, make, and sell goods can form the basis for regulating and ensuring sustainability initiatives.

The Role of Government and Public Policy

The issue of cross-border governance must deal with the fact of national sovereignty. When we think about global sustainability, all of the local and nation-state components of global sustainability are subject to policy direction from within sovereign nations. That is clear and works well. However, environmental and sustainability issues that cross borders must deal with the constraints of national sovereignty. I do not expect to see the force of national sovereignty fade in the near future, and thus global-scale sustainability issues must utilize those policy instruments now in place for dealing with international issues. The international regulation of trade, environment, arms, communications, and even postal delivery has a long and even productive history. Nation-states, like people and companies, are capable of making agreements that advance their mutual interests.

Government is a central player in establishing and maintaining sustainability at the community, city, state, national, and international level. The idea that an unregulated market can achieve sustainability is absurd. Both government and the private sector must work together to advance economic development with the least possible impact. The problem, of course, is that, with apologies to my friends at the United Nations, there is no global government. Public policy about global sustainability requires international diplomacy and the committed actions of sovereign nation-states. While a world court and other mechanisms of adjudication and enforcement of international law have evolved, they do not have the force of law that can be enforced within sovereign nations.

Nevertheless, the scale of economic development and organization is now firmly global, as is the media and communications. Culture has begun to follow, and so governance and public policy cannot be too far behind. The form of governance often, but not always, adheres to its functional needs. My own expectation is that the real action in sustainability public policy will begin at the community level, grow to the municipal level, and then be reflected in ever larger jurisdictions. As the late Tip O'Neil once said: "All politics is local." That includes the politics of sustainability policy.

Chapter 8

Conclusions

This volume has attempted to present an introduction to the field of sustainability management. I have tried in these few pages to begin the process of combining the fields of environmental policy and management with the fields of business management and finance. The old notion of environmental policy is that it is an add-on to a modern economy. It makes the economy more civilized and pleasant, but in some ways it is a frill, a luxury. The field of sustainability management is based on different premises. It assumes that the modern economy on a crowded planet requires that environmental issues be factored into general issues of production and consumption and that our wealth depends on a well-functioning ecosphere. You cannot trade off environment and economic growth—the two are interconnected.

To conclude, I will return to the basics and ask if sustainability management is a feasible concept and field in the real world. This requires that we address three central questions:

1. Can we manage this complexity, or will the modern economy destroy this planet?
2. What are the key technical, financial, organizational and political challenges we face?
3. What are the odds of success?

Can We Manage This Complexity, or Will the Modern Economy Destroy This Planet?

No one can predict the future, but we do need to try to project forward to address this question. The modern economy is now truly global. The computer I am writing this on is made up of technologies and parts designed and manufactured in dozens if not hundreds of locations. The environmental impact of this computer through its entire lifecycle falls under many jurisdictions, and the process of producing this product is a complex task drawing on many technologies and organizational forms. Multiply these transactions by all of the products in your own household and you have a sense of the complexity of the task of managing sustainability.

The planet is under enormous stress as the population increases and new technologies are introduced. When local organizations deploy technologies that poison the air, water, or land, local governments have shown the capacity to regulate that behavior. After forty years of environmental protection in the United States, Japan, and Western Europe, all of these places are less polluted today than they were in 1970. This is true even though the population of the world and its economies have expanded dramatically. This is the strongest indication that sustainability management may very well be feasible. We have decoupled the growth of the GDP and the growth of pollution.

Of course, that is in the developed world. In many respects, the technological base for development in newly developed nations is far more toxic than the one that was present when the United States developed. Industrialization in America did not have the benefit or the costs of the advanced plastics and other chemicals now in use. We also did not have a more developed economic model elsewhere in the world to compare ourselves to, pushing us to develop faster than we might have been ready to move. Finally, we did not have the distorting influence of the global media or potentially intrusive international institutions such as the IMF and the World Bank.

We have demonstrated an ability to control the worst effects of industrialization and urbanization. We are less confident in our ability to control the more subtle or indirect effects. One of our greatest deficiencies is in our ability to measure and understand the overall health of the biosphere. Many critical resources are not measured often enough and accurately enough to be managed. With the advances in remote sensing, satellite imaging, computer and communication technology, and environmental sciences generally, it is reasonable to assume that we will continue to improve our ability to measure and understand the planet's ecological functioning.

If we understand the impacts of technology, we can develop safer technologies as they are needed and moderate our use and ease the impact of the ones we must retain. In many respects, global climate change will be a test of our ability to stay ahead of the impact of our technology. We understand the impact of fossil fuels on our atmosphere and we are starting to develop the control technologies and public policies needed to reduce those impacts. Nevertheless, it is unclear whether or not we will be able to respond effectively to this global sustainability management problem.

I should note that I do not believe that the most likely outcome of the challenge to sustainability to be the destruction of the planet's capacity to sustain human life. If we fail to sustain natural ecosystems and reduce global warming, we will develop technological means to purify air and water and grow food. We will slowly erode the resources we now depend on and replace them with human-engineered substitutes. A person with twenty-first-century perceptions and values might not like or recognize the place, but we are not likely to experience the drama of ecological disaster movies. In many respects, when we argue for sustainability we are making a very conservative argument to preserve the oceans, meadows, and mountains that we now know and enjoy. While we cannot maintain the quality of life we now have or provide *this* lifestyle to people in the developing world if we don't preserve the ecosystem's productive capacity, I do not believe that the survival of our species is at stake.

Although we do not have the technology today to exist without natural systems, someday we will develop those technologies. In fact, the destruction of those systems will likely hasten the development of replacements. That is why at its heart sustainability management is about values and ethics. We seek to sustain the ecosphere out of our belief that it is our obligation to do so—that to destroy the ecosphere is ethically and morally wrong and to preserve it is ethically and morally correct.

In sum, the modern economy will never destroy the planet, but it will impair it. It already has damaged many natural resources. The question is how much destruction we will allow and at what point we will direct more of our ingenuity to preserving ecosystems than to destroying them. In the final analysis, that is a value choice as well. As a policy analyst and a student of history, I would not minimize the potential impact of values and ideas on human behavior. The lifestyle we now enjoy is a direct result of ideas that were generated during the Enlightenment, if not earlier. My underlying optimism is based on the human capacity to change and on the very fact that you are reading this book.

What Are the Key Technical, Financial, Organizational, and Political Challenges That We Face?

In my view, the key technological challenge is energy. If we could inexpensively and safely harness the energy of the sun or of the atom and provide it at low cost and low environmental impact throughout the economy, we would vastly increase our ability to transform biochemical matter and manufacture air, food, and water. This would enable us to maintain current ecosystems by recycling and reusing the resources we now utilize instead of destroying those that are still relatively removed from human impacts.

In addition to a less destructive and lower-cost form of energy, the other technical challenge is simply to develop a better understanding of the state of the planet: the functioning of its ecosystems and the impact of humans on their functioning. Our capacity for Earth observation and measurement must be improved. The science of the environment and a deeper understanding of life on this planet must also be improved. While our knowledge continues to advance rapidly, this process needs to be accelerated to keep pace with the human-induced changes that are now underway.

The key financial challenge is funding the capital cost of the transition to green infrastructure. Capital seeks the highest rate of return, and in order to attract investment in this early stage of technology development, government incentives and regulation are needed. Once these investments are made, the costs of energy, water, and transport are reduced, along with their environmental impacts. However, the environmental benefits are not monetized and so they are not typically part of the calculus of investment. The challenge of attracting capital and enhancing the rate of return will change as technology develops and expands in scale. These factors reduce the cost side of the cost-benefit equation. Regulation can be used to require investments that reduce an organization's environmental impact. If these rules are credible, they can convince management that the capital costs of improvement are a necessary cost of doing business. This will then influence internal investment decisions and external investment in the development of new production or pollution control equipment.

We have discussed the issue of organizational capacity before, but if this organizational capacity is to be developed, the field of sustainability management must replace the field of general management. As I mentioned earlier, just as the development of modern accounting systems and staffing by accountants became common practice in the corporate world during the twentieth century, I expect similar practices to be developed for sustainability. While it may not get to the point of certification, some type of sustainability professional will become commonplace in the organizations of the twenty-first century.

The politics of sustainability are in many respects the final piece of the puzzle. While the economic facts of a finite planet will push corporations toward sustainability practices, rules of the game must be put in place to ensure a level playing field. The temptation to grab short-term gains at the expense of the environment must be avoided. Regulations, however, must be established with attention paid to the full impact of the rules. For example, sustainability rules need to minimize their impact on small businesses, or the only remaining firms will be large ones with the resources to comply. In some cases, the costs of compliance can favor large corporations over small ones. However, small firms can sometimes create and house creative new enterprises, so it is important to protect them.

Politics often includes symbolic dimensions, and some of the dangers of sustainability politics became apparent during the debate over climate policy in the U.S. Congress in 2009 and 2010. States with large coal mining and oil refining businesses were bombarded with media and lobbying messages sent by those businesses. These in turn were reflected in the national debate. Interest groups tended to dominate and distort the debate, even to the extent of questioning fundamental scientific fact in their efforts to undermine the formulation of climate policy.

The politics of sustainability tends to be less contentious when the issues are local and the impacts are short term and tangible. A local water shortage generates a very different political battle than a global climate issue. Issues related to short-term economic benefits will tend to dominate these political conflicts. Still, the ability of interest groups seeking immediate economic gains to dominate political dialogue should not be underestimated. Scientific fact and even economic interests can be cast aside in the wake of the propaganda of a determined interest group.

The role of climate "deniers" in the debate over climate policy is a case in point, but many other examples can easily be found. In New York State in early 2010, a proposed tax on sweetened beverages resulted in a relentless barrage of radio spots warning of the massive economic impact of this tax should it be enacted. In 2010, the debate over national health care was dominated by opponents willing to say just about anything to defeat the proposal. These misconceptions are magnified by the 24/7 media in an endless cycle of bad information. Eventually, the average voter internalizes these absurd messages, such as the growth of the ice caps, the president's false birth certificate, and "death panels" deciding when to "pull the plug on grandma." These messages can gain currency for a while, but my clear sense is that eventually the truth gets out and reality reasserts itself. Still, it can take a long time for that to happen, and we can expect sustainability politics to be dominated by propaganda with some frequency. It is difficult to communicate complex and even subtle messages about long-term impacts under these conditions.

What Are the Odds of Success?

A careful analysis of the crisis of global sustainability might lead you to believe that the planet's chances of survival are slim. However, I wouldn't base our judgment of the probability of success on an analysis of current data alone. The human species has shown enormous resiliency and creativity. Malthus, of course, is the cautionary tale here. His analysis indicated that due to overpopulation we would eventually run out of food. His brilliant and careful math simply left technological innovation out of the equation. In our own time, technology still cannot be predicted, nor can the degree of social learning now underway. The issue of environmental sustainability has vastly increased in currency over the past half century. People are more aware of the constraints placed on their activity by the environment and are more aware of the need to be careful in their use of natural resources and settings.

A concern for the future and posterity is never assured, but human civilization has evidenced the ability to maintain the treasures of the past while preserving them for generations to come. At the same time, conflict, terrorism, and mass destruction always lurk in the shadows or over the next horizon. Carelessness and thoughtlessness are also human traits. Nevertheless, I am an optimist, and so I believe that although we will face setbacks and even catastrophes, we will learn from them. Despite the presence of nuclear weapons all over the world, we have managed to go over six decades without using one in warfare. I suspect we will damage the planet more before we learn how to sustain it, but someday we will get it together and develop the expertise needed to sustain this planet and lift people out of poverty and misery.

In the end, I do not think that my belief will hold up to an analytic and data-based argument, but I will hold on to that belief just the same. My faith in humanity seems to be hardwired into my psyche and resists the evidence to the contrary. I am certainly aware of history and of Nazis, terrorists, and today's contemporary genocidal maniacs along with all of the evil and sick people in world. Nevertheless, I believe that they are outnumbered by the vast number of good and decent people who want to live in peace with their neighbors and see their children—and all children—live up to their potential. I could be wrong, but I am betting that someday all high-quality organizational management will be sustainability management.

The younger generations will carry the sustainability load on their shoulders. The following case study, based on a blog post I wrote for the *Huffington Post*, explores future generations' potentials for sustainability management. If there is hope anywhere, it is in the future generations.

CASE STUDY: THE SUSTAINABILITY GENERATION COMES OF AGE

The political noise of self-aggrandizing interest groups and people willfully dismissing the science of sustainability is receding in the face of the energy, intelligence, and common sense of the students that I work with here at Columbia University. The views of this generation of students are not a reflection of some giant left-wing conspiracy to propagandize young people in the nation's K–12 education system. They are a reflection of the reality of modern life. When I was growing up in the 1960s, our planet had about three billion people. Today, we have more than twice that number. This nation has grown by over one hundred million people during that time. While this is a big country and an even bigger planet, everyone knows of a place that they hiked to or camped at as a kid that's now a strip mall or a condo. You don't need to be a scientist to know that while the population and material consumption are growing, the size of the planet remains the same. That's common sense.

It's true that technology and better management can help us enhance the planet's ability to sustain its human population. The students I work with know that and are eager to work on those solutions. They do not see the future as one in which they are sitting alone and hungry in the dark. Their worldview is built upon a deep understanding of the challenges of global sustainability and represents both a generational and cultural shift that has accelerated in the twenty-first century. This worldview is widespread in this country and is even growing in places like China, despite state-supported efforts to suppress it.

I think that the increased interest in sustainability comes from a sense that the world is getting more complicated and that the future belongs to those who study it, understand it, and learn how to manage it. One of the dangers of the modern world is the simple fact that the global economy, the 24/7 media, and the biosphere can be hard to understand. Lots of folks make their living hawking oversimplified answers to issues they don't really understand. Climate science is a good example. There is no question that the planet's atmosphere is getting warmer and that the impact of nearly seven billion people and their fuels is real and substantial. What we don't know is what effect these realities will have on human settlements and the planet's ecosystems. We have a lot to learn and more to study, and we can easily get into trouble when we confidently predict future effects as if they are certainties and not simply probabilities.

As an educator, I believe in the power of knowledge and in the importance of humility in the face of uncertainty. As a public policy analyst,

(*continued*)

CASE STUDY: THE SUSTAINABILITY GENERATION COMES OF AGE (*continued*)

I know that action cannot wait until we achieve perfect understanding. We never achieve perfect understanding. Government and public policy never really solve problems; they simply make them less bad. We will never end poverty, crime, or environmental degradation; our real goal is simply to reduce these evils. In the case of sustainability management, we need to teach our students how to understand and draw upon diverse fields of knowledge. We need to learn how to draw on real expertise and how to reduce the temptation to impose ideological interpretations on empirical data.

I believe that over the next decade or so, all management will become sustainability management. In other words, it won't be enough to simply know about organizations, strategy, marketing, and finance. The people who run our institutions and industries will need to know about energy efficiency, waste management, and environmental impacts as well.

Student interest in these issues is matched by interest in what those of us on college campuses sometimes call the "real world." The "green jobs" market is growing as the economy begins the long transformation into what many of us hope will be a more sustainable form of economic growth. The cultural shift I see in the current generation of students is having an influence on the media, politics, the economy, and even that slow-moving, slow-changing place called the American university. Our society's powerful institutions know that they must respond to this impulse toward environmental sustainability. They know this because they are themselves being transformed from within. As this sustainability generation takes its place in business, government, and academia, the pace of change will accelerate. And that more than anything else provides hope for the future.

References

ACFNewsource. 2002. Reduce, reuse, recycle comes to life. ACFNewsource. http://www.acfnewsource.org/environment/calmax.html.

Adams, M. 2005. Rising popularity of medical tourism reveals deterioration of US healthcare system. *NaturalNews.com*. http://www.naturalnews.com/007097.html.

Added Value. 2009. History. http://www.added-value.org/history.

Allen, A. 2001. Containment landfills: The myth of sustainability. *Engineering Geology* 60, nos. 1–4: 3–19.

Associated Press. 2006. Farms waste much of world's water. *Wired* (March 19).

———. 2009. Price drop slows farm economy. *Topeka Capital-Journal* (April 2).

Aston, A. 2008. Energy Star doesn't mean your fridge is green: Critics of the energy-efficiency rating system say companies such as Samsung and LG are gaming it. *Business Week* (October 2).

Bartlett, B. 2009. Tax tea party time? *Forbes.com*. http://www.forbes.com/2009/04/09/tea-party-taxes-opinions-columnists-bartlett.html.

Baue, W. 2004. Zero waste and industrial ecology: A talk with Xerox VP for Environment, Health, and Safety Jack Azar. *SocialFunds.com: Sustainability Investment News* (November 11).

Black, R. 2004. Study finds benefits in GM crops. *BBC News* (November 29).

Box, D. 2003. Greening cities. *The Ecologist* 33, no. 3: 56–57.

Bradsher, K. 2009. China racing ahead of U.S. in drive to go solar. *New York Times* (August 24).

Braybrooke, D., and C. Lindblom. 1963. *A strategy of decision*. New York: Free Press of Glencoe.

Broder, J. M., and M. Connelly. 2010. Poll finds deep concern about energy and economy. *New York Times* (June 21).

California Department of Resources Recycling and Recovery. 2009. CalMax. http://www.calrecycle.ca.gov.

CPUC = California Public Utilities Commission. 2009. Program funding. http://www.cpuc.ca.gov/PUC/energy/Energy+Efficiency/EE+General+Info/ee_funding.htm.

Campbell, S. 1996. Green cities, growing cities, just cities? Urban planning and the contradictions of sustainable development. *Journal of the American Planning Association* 62, no. 3: 296–312.

Center for Disease Control. 2010. Sources of lead: Water. http://www.cdc.gov/nceh/lead.

Center for Sustainable Energy. 2010. Property assessed clean energy programs. http://energycenter.org/index.php/public-affairs/property-assessed-clean-energy-pace.

Central Intelligence Agency. 2010. Country comparison: Military expenditures. https://www.cia.gov/library/publications/the-world-factbook/rankorder/2034rank.html.

Certified Naturally Grown. 2010. http://www.naturallygrown.org.

Choi, C. Q. 2010. New poll shows support for use of renewable energy like wind, solar. *Live Science* (June 10): 23.

Christo and Jeanne-Claude. 2010. The Gates. http://www.christojeanneclaude.net/tg.shtml.

City of Phoenix. 2010. ASU downtown Phoenix campus. http://phoenix.gov/downtown/asucop3.html.

Cohen, B. 2006. Urbanization in developing countries: Current trends, future projections, and key challenges for sustainability. *Technology in Society* 28: 63–80.

Cohen, S. 2001. A strategic framework for devolving responsibility and functions from government to the private sector. *Public Administration Review* 61, no. 4: 432–440.

———. 2008. Wasted: New York's giant garbage problem. *New York Observer.* http://www.observer.com/2008/wasted-new-york-citys-giant-garbage-problem.

———. 2009a. Developing a sustainable planet: The basics. *New York Observer.* http://www.observer.com/2009/developing-sustainable-planet-basics.

———. 2009b. Sustainable financing for mass transit in New York. *Huffington Post.* http://www.huffingtonpost.com/steven-cohen/sustainable-financing-for_b_408266.html.

———. 2010a. Changing Obama's management style alone will not prevent the next environmental catastrophe. *Huffington Post.* http://www.huffingtonpost.com/steven-cohen/changing-obamas-managemen_b_602787.html.

———. 2010b. "Finding the Cash for Sustainable Energy." *The Huffington Post.* Retrieved on August 26, 2010, from http://www.huffingtonpost.com/steven-cohen/finding-the-cash-for-sust_b_494767.html.

———. 2010c. The gulf spill and effective management of regulation. *Huffington Post.* http://www.huffingtonpost.com/steven-cohen/the-gulf-spill-and-effect_b_576522.html.

Cohen, S., and W. Eimike. 2008. *The responsible contract manager.* Washington, D.C.: Georgetown University Press.

Cohen, S., and S. Schonhardt. 2009. Bring green principles into the American economy. *New York Observer* (April 2). http://www.observer.com/2009/green/bringing-green-principles-american-economy.

Colcanis, P. 2003. Back to the future: The globalization of agriculture in historical context. *SAIS Review* 23, no. 1.

Common, M., and S. Stagl. 2005. *Ecological economics.* Cambridge: Cambridge University Press.

ConEd.com. 2010. How Con Edison is addressing New York's energy and environmental issues. http://www.coned.com/energyny.

ConsumerReports.com. 2008. Energy Star has lost some luster: The program saves energy but hasn't kept up with the times. *Consumer Reports* (September).

CENY = Council on the Environment of New York City. 2009. Why shop at greenmarket? http://www.cenyc.org/greenmarket/whygreenmarket.

Crosson, P. 1986. Sustainable food production: Interactions among natural resources, technology and institutions. *Food Policy* 11, no. 2: 143–156.

Crow, M. 2010. Aggressive pursuit of economic competitiveness key to long-term prosperity. *Arizona Republic.* http://www.azcentral.com/arizonarepublic/opinions/arizona2020/articles/2010/01/22/20100122crow24-vision.html.

DePalma, A. 2007. City's Catskill water gets 10-year approval. *New York Times* (April 13).

Department of Agriculture and Markets, State of New York. 2005. Survey says wholesale markets good for farmers, consumers. *Department of Agriculture and Markets News.* http://www.agmkt.state.ny.us/AD/release.asp?ReleaseID=1403.

Department of City Planning, City of New York. 2010. Primary land use table. http://www.nyc.gov/html/dcp/pdf/landusefacts/landuse_tables.pdf.

Department of Environmental Protection, City of New York. 2006. New York City 2006 water supply and quality report. http://www.nyc.gov/html/dep/pdf/wsstate06.pdf.

———. 2007. New York City's wastewater treatment system. http://www.nyc.gov/html/dep/pdf/wwsystem.pdf.

———. 2010. Upstate watersheds. http://home2.nyc.gov/html/dep/html/watershed_protection/html/statement.html.

Department of Parks and Recreation, City of New York. About GreenThumb. http://www.greenthumbnyc.org/about.html.

Derraik, J. 2002. The pollution of the marine environment by plastic debris: A review. *Marine Pollution Bulletin* 44, no. 9: 842–852.

Douglas, E. 2008. There's gold in them there landfills. *New Scientist.* http://www.newscientist.com/article/mg20026761.500-theres-gold-in-them-there-landfills.html.

Dunlea, M., S. McCarthy, and S. Pasquanton. 2005. A community food security agenda for New York. Hunger Action of New York State Web site. http://www.hungeractionnys.org/CommunityFoodPaper.pdf.

Earth Institute. Innovative methods of green energy finance panel discussion. Columbia University. http://www.earth.columbia.edu/videos/watch/192.

Eaton, C., and A. Shepherd. 2001. Contract farming: Partnerships for growth. *FAO Agricultural Services Bulletin* 145.

Eccleston, P. 2008. EU pesticide rules would decimate crop yields. *Telegraph* (June 22).

Edwards, C. 1989. The importance of integration in sustainable agricultural systems. *Agriculture, Ecosystems and Environment* 27: 25–35.

Ehrenfeld, J. 2005. The roots of sustainability. *MIT Sloan Management Review* 46, no. 2: 23–25.

Energy Division, City of Portland. 2001. Energy efficiency success story: LED traffic signals = energy savings. http://www.portlandonline.com/shared/cfm/image.cfm?id=111737.

EIA = Energy Information Administration, U.S. Department of Energy. 2008. Energy consumption, expenditures, and emissions indicators, 1948–2008. http://www.eia.doe.gov/emeu/aer/txt/ptb0105.html.

Environmental Leader. 2009. Sustainability initiatives cut costs by 6–10%. *Environmental Leader: The Executive's Daily Briefing* (June 9).

FHFA = Federal Housing Finance Agency. 2010. FHFA statement on certain energies retrofit loan programs. http://www.fhfa.gov/webfiles/15884/PACESTMT7610.pdf.

Fiedler, M. et al. 2009. Energy management in New York City's public housing. NYCHA. http://www.columbia.edu/cu/mpaenvironment/pages/projects/spring09/EneManNYCPubHousReport.pdf.

Findlaw.com. 2010. N.Y. AGM. LAW § 259: NY Code—Section 259: Legislative findings. http://codes.lp.findlaw.com/nycode/AGM/22/259.

Foodworks New York. 2009. Memo to City Council, December 9.

Foreschle, L. 1999. Environmental assessment and specification of green building materials. CalRecycle Web site. http://www.calrecycle.ca.gov/GreenBuilding/Materials/CSIArticle.pdf.

Gadgil, A. 1998. Drinking water in developing countries. *Annual Review of Energy and the Environment* 23: 253–286.

Gallopoulos, N. 2006. Industrial ecology: An overview. *Progress in Industrial Ecology: An International Journal* 3, nos. 1–2: 10–27.

Garfield, S., C. Walker, and Y. Nelson. 2007. Integrated energy policy report. California Energy Commission Web site. http://www.energy.ca.gov/2007publications/CEC-100-2007-008/CEC-100-2007-008-CMF.PDF.

GIT = Georgia Institute of Technology, Capgemini, Oracle, DHL. 2008. Third-party logistics providers play key role in integration, sustainability, and security of the supply chain. http://www.logisticshandling.com/absolutenm/templates/article-transport_distribution.aspx?articleid=330&zoneid=23.

Gotham Gazette. 2007. The garbage glut. *Gotham Gazette.* http://www.gothamgazette.com/article/iotw/20071113/200/2343.

Grynbaum, M. 2009. Drastic cuts are expected as MTA unveils budget. *New York Times.* http://www.nytimes.com/2009/12/14/nyregion/14mta.html.

Hagan, K. 2009. Public transportation and energy efficiency initiatives. *Associated Content.* http://www.associatedcontent.com/article/2095121/new_york_the_most_energyefficient_city.html.

Hamilton, R. 2009. Agriculture's sustainable future: Breeding better crops." *Scientific American* (June).

Hanano, J., and C. Baanante. 2006. Agricultural production and soil nutrient mining in Africa: Implications for resource conservation and policy development—Summary. *IFDC.* http://www.cababstractsplus.org/abstracts/Abstract.aspx?AcNo=20063142849.

Handley, G., and L. Larsen. 2007. The radiant astonishment of existence: Two interviews with Marilynne Robinson, March 20, 2007, and February 9, 2007. *Literature and Belief* 27, no. 2.

Hargreaves, S. 2009. Cash for caulkers could seal $12,000 a home. *CNNMoney.com* (December 8).

Harris, T. 2002. How light emitting diodes work. http://electronics.howstuffworks.com/led.htm.

Hellström, D., U. Jeppsson, and E. Kärrman. 2000. Assessment methodologies for urban infrastructure. *Environmental Impact Assessment Review* 20, no. 3: 311–321.

Hoover, K. 2009. Business bankruptcies rise 64 percent. *Kansas City Business Journal.* http://www.bizjournals.com/kansascity/stories/2009/08/24/daily17.html.

Hopkins, M. 2009. Sustainability, but for managers. *MIT Sloan Management Review* 50, no. 3: 10–16.

Hopper, N., C. Goldman, D. Gilligan, T. Singer, and D. Birr. 2007. A survey of the U.S. ESCO industry: Market growth and development from 2000 to 2006. Lawrence Berkeley National Laboratory Web site. http://eetd.lbl.gov/ea/EMP/reports/62679.pdf.

Jusko, J. 2009. Financing energy projects. *Industryweek* (June 17).

Just Food. 2010a. About us. http://www.justfood.org/about-us.

———. 2010b. Community chef training. http://www.justfood.org/community-food-education/community-chef-training.

Kaebernick, H., and K. Sami. 2006. *Environmentally sustainable manufacturing: A survey on industry practices.* Sydney: University of New South Wales Life Cycle Engineering and Management Research Group.

Kahlown, M., M. Azam, and W. Kemper. 2006. Soil management strategies for rice-wheat rotations in Pakistan's Punjab. *Journal of Soil and Water Conservation* 61, no. 1: 40–44.

Kanter, J. 2010. EU signals big shift on genetically modified crops. *New York Times* (May 9).

Kaupp, A. 2008. Article #18: Energy auditors under the EC act: Loved, redundant or a nuisance? http://www.energymanagertraining.com/kaupp/Article18.pdf.

Kennedy, J. 2008 [1963]. 1963 commencement. http://www1.media.american.edu/speeches/Kennedy.htm.

Kirby, A. 1999. Friend or foe? *BBC News: Special Report* (April 6).

Lall, U., T. Heikkila, C. Brown, and T. Seigfried. 2008. Water in the 21st century: Defining the elements of global crises and potential solutions. *Journal of International Affairs* 61, no. 2: 1–17.

LaMonica, M. 2009. Google crashes smart grid party. *CNET News* (February 10). http://news.cnet.com/google-crashes-the-smart-grid-party.

La Vina, A., L. Fransen, P. Faeth, and Y. Kurauchi. 2006. Reforming agricultural subsidies: "No regrets" policies for livelihoods and the environment. World Resources Institute Web site. http://pdf.wri.org/reforming_ag_subsidies.pdf.

Layzer, J. (Speaker). 2008. Sustaining cities: Environment, economic development, and empowerment. MIT World Web site. http://mitworld.mit.edu/video/574.

———. [Interview with Michael Hopkins] 2009. An urban planner's dream. *MIT Sloan Management Review.* http://sloanreview.mit.edu/the-magazine/files/pdfs/50430SxW.pdf.

Ledbetter, Anita. 2009. Rethinking reuse: How incorporating the basic principals of reused materials and techniques into building design positively impacts the environment. Build San Antonio Green Web site. http://buildsagreen.org/articles/Rethinking-Reuse.html.

Lee, E., and J. Schwab. 2005. Deficiencies in drinking water distribution systems in developing countries. *Journal of Water and Health* 3, no. 2: 109.

Lento, B. 2009. Leap-frogging to smart grid success: Lessons from the telecommunications industry. *Smart Grid News* (December 8).

Ling, K. 2009. Senate panel to study details of setting up electric 'smart grid.' *New York Times* (March 9).

Litos Strategic Communication. 2008. Smart grid stakeholder book: Environmental groups. U.S. Department of Energy Web site. http://www.oe.energy.gov/DocumentsandMedia/Environmentalgroups.pdf.

Living Memorials Project. 2003. http://www.livingmemorialsproject.net/toolbox/people/NARRATIVES/communitygarden.htm.

Locke, R. [Interview with Michael Hopkins]. 2009. Sustainability as fabric—and why smart managers will capitalize first. *MIT Sloan Management Review.* http://sloanreview.mit.edu/beyond-green/sustainability-as-fabric-and-why-smart-managers-will-capitalize-first.

Lohr, S. 2008. Digital tools help users save energy, study finds. *New York Times* (January 10).

Lovins, A. 2005. More profit with less carbon. *Scientific American.* http://www.scientificamerican.com/article.cfm?id=more-profit-with-less-car.

Lowenstein, R. 2006. Who needs the mortgage-interest deduction? *New York Times* (March 5).

Lubber, M. 2008. Short-term strategies don't work for Wall Street or the planet. *Harvard Business Review* (September 24). http://blogs.hbr.org/leadinggreen/2008/09/shortterm-strategies-dont-work.html.

Matson, P. A., et al. 1997. Agricultural intensification and ecosystem properties. *Science* 277, no. 5325: 504–509.

McGillis, B., and A. McDonald. 2009. Pew finds clean energy economy generates significant job growth. The Pew Charitable Trusts Web site. http://www.pewtrusts.org/news_room_detail.aspx?id=53254.

McHarg, I. 1982 [1967]. *Design with nature.* New York: John Wiley & Sons, Inc.

McReynolds, K., S. Pater, and K. Uhlman. 2005. Watershed basics. Master Watershed Steward program Web site. http://ag.arizona.edu/watershedsteward/resources/docs/guide/(8)Hydrology.pdf.

Melwani, L. Underground tales. *Little India.* http://www.littleindia.com/news/142/ARTICLE/1551/2004-05-05.html.

Mendonca, Miguel. 2007. Feed-in tariffs: Accelerating the deployment of renewable energy. Hamburg: World Future Council. http://www.earthscan.co.uk/Portals/0/Files/Sample%20Chapters/9781844074662.pdf.

Metropolitan Transit Association, City of New York. 2007. Subway centennial. http://www.transitmuseumeducation.org/100/centennial.html.

MIT Sloan Management Review, in collaboration with Boston Consulting Group. 2009. Thriving in the sustainability economy: Capabilities required. *MIT Sloan Management Review.* http://sloanreview.mit.edu/the-magazine/files/2009/03/recon-960-1052.jpg.

More Gardens! 2010. History. http://www.moregardens.org/node/141.

Moreno, E., and R. Warah. 2006. The state of the world's cities report 2006/7: Urban and slum trends in the 21st century. *UN Chronicle* 43, no. 2. http://un.org/Pubs/chronicle/2006/issue2/0206p24.htm.

Muller, J. 2009. Back to business at GM. *Forbes Magazine* (August 3).

Municipal Water Finance Authority, City of New York. 2010. About us. http://www.nyc.gov/html/nyw/html/aboutus.html.

NAESCO.org. 2009. What is an ESCO? National Association of Energy Service Companies Web site. http://www.naesco.org/ resources/esco.htm.

NSAIS = National Sustainable Agriculture Information Service. Reducing food miles. http://attra.ncat.org/farm_energy/food_miles.html.

Navarro, M. 2009. Bloomberg drops an effort to cut building energy use. *New York Times* (December 4).

Neukrug, H. (Director, Office of Watersheds for the Philadelphia Water Department). 2001. Drinking water needs and infrastructure. Testimony before Environment and Hazardous Materials Subcommittee of the Committee on Energy and Commerce, U.S. House of Representatives. http://www.win-water.org/legislativecenter/neukrug3_28_01.shtml.

New Jersey Clean Energy Program. 2009. Commercial, industrial and local government. New Jersey Board of Public Utilities Web site. http://www.njcleanenergy.com/commercial-industrial/programs/local-government-energy-audit/steps-participation.

New York State Division of Criminal Justice Services. 2008. Index crimes reported to police by region, 1998–2008. http://criminaljustice.state.ny.us/crimnet/ojsa/indexcrimes/regiontotals.pdf.

Niblack, P. Cost comparison: The New York City Independent Budget Office says more recycling could help lower the city's trash costs. *Entrepreneur*. http://www.entrepreneur.com/tradejournals/article/166944606.html.

North, D. 2009. Landfill or incineration: Does increased incineration mean less recycling? *MyZeroWaste.com*. http://myzerowaste.com/2009/12/landfill-or-incineration-does-increased-incineration-mean-less-recycling.

Northwestern University Library. 2009. New released: 2007 census of agriculture. *Government Information News* (March 18). http://www.library.northwestern.edu/govinfo/news/2009/03/new_released_2007_census_of_ag.html.

O'Carroll, P., ed. 2003. *Public health informative and information systems*. New York: Springer.

Odum, H., and E. Odum. 1981. *Energy basis for man and nature*. New York: McGraw-Hill.

Office of the Comptroller, City of New York. 2009. Audit on the Department of Environmental Protection's progress in constructing the Croton water treatment plant. http://www.comptroller.nyc.gov/BUREAUS/AUDIT/09-01-09_FR08-121A.shtm.

Office of the Mayor, City of New York. 2005. Mayor Michael Bloomberg announces $254 million economic impact of *The Gates* on New York City. http://www.nyc.gov/portal/site/nycgov/menuitem.c0935b9a57bb4ef3daf2f1c701c789a0/index.jsp?pageID=mayor_press_release&catID=1194&doc_name=http%3A%2F%2Fwww.nyc.gov%2Fhtml%2Fom

%2Fhtml%2F2005a%2Fpr078-05.html&cc=unused1978&rc=1194&ndi=1.

———. 2010. PlaNYC 2030: Energy. http://www.nyc.gov/html/planyc2030/html/plan/energy.shtml.

Organic Seeds Trust. 2010. Certified naturally grown. http://www.seedstrust.com/org/organic.html.

Organic Trade Association (OTA). 2010. http://www.ota.com/index.html.

OECD = Organization for Economic Cooperation and Development. 2009. Sustainable manufacturing and eco-innovation: Framework, practices, and measurement: Synthesis report. http://www.oecd.org/dataoecd/15/58/43423689.pdf.

Organization for Economic Cooperation and Development/Nuclear Energy Agency. 2010. Projected costs of generating electricity: 2010. *Source OECD* 2010, no. 3: 1–218.

Pace Financing. 2010. http://pacefinancing.org.

Painuly, J. 2001. Barriers to renewable energy penetration: A framework for analysis. *Renewable Energy* 24, no. 1: 73–89.

Pant, M., A. Sharma, and P. Sharma. 1980. Evidence for the increased eutrophication of Lake Nainital as a result of human interference. *Environmental Pollution* (Series B) 1: 149–161.

Peterson, G., S. Cunningham, L. Deutsch, J. Erickson, A. Quinlan, E. Raez-Luna, R. Tinch, M. Troell, P. Woodbury, and S. Zens. 2000. The risks and benefits of genetically modified crops: A multidisciplinary perspective. *Ecology and Society* 4, no. 1.

Pew Center for Global Climate Change. 2009. Decoupling in detail. http://www.pewclimate.org/docUploads/Revenue_Decoupling_detail_0.pdf.

Pew Charitable Trusts. 2009. *Clean energy economy: Repowering jobs, businesses, and investments across America*. Washington, D.C.: Pew Charitable Trusts with Collaborative Economics, Inc. http://www.pewcenteronthestates.org/uploadedFiles/Clean_Economy_Report_Web.pdf.

Pew Commission on Industrial Farm Animal Production 2008. Putting meat on the table: Industrial farm animal production in America. http://www.ncifap.org/reports.

Pimm, S., and P. Raven. 2000. Biodiversity: Extinction by numbers. *Nature* 403 (February 24). http://www.nature.com/nature/journal/v403/n6772/full/403843a0.html.

Rampell, C. 2009. Savings rate rising toward mediocrity. *New York Times*. http://economix.blogs.nytimes.com/2009/06/26/savings-rates-rising-toward-mediocrity.

Rasool, R., S. Kukal, and G. Hira. 2007. Soil physical fertility and crop performance as affected by long term application of FYM and inorganic fertilizers in rice-wheat system. *Soil & Tillage Research* 96: 64–72.

Reganold, J., R. Papendick, and J. Parr. 1990. Sustainable agriculture. *Scientific American* 262: 112–120.

Renewable Energy Focus. 2010. Wind and solar power cost sensitive to load factors. *RenewableEnergyFocus.com*. http://www.renewableenergyfocus.com/view/8359/wind-and-solar-power-cost-sensitive-to-load-factors.

Rijsberman, F. 2006. Water scarcity: Fact or fiction? *Agricultural Water Management* 80, nos. 1–3: 5–22.

Robinson, A., and D. Schroeder. 2009. Greener and cheaper: The conventional wisdom is that a company's costs rise as its environmental impact falls; Think again. *Wall Street Journal* (March 23).

Royte, E. 2007. On the water front. *New York Times* (February 17). http://www.nytimes.com/2007/02/18/nyregion/thecity/18feat.html.

Sachs, J. 2005. *The end of poverty*. New York: Penguin.

Saijo, E. 2005. Redesigning the office copier—One manufacturer's efforts to conserve resources. *Mail*. http://www.japanfs.org/en/mailmagazine/newsletter/pages/027881.html.

Samadpour, M., P. Evans, K. Everett, P. Ghatpande, G. Ma, and R. Miksch. 2005. Occupational and environmental exposure to waterborne microbial diseases. In *Textbook of Clinical Occupational and Environmental Medicine*, 2nd ed., 1195–1221. New York: Elsevier.

Sanderson, E., and M. Boyer. 2009. *Mannahatta: A natural history of New York City*. New York: Abrams.

Satterthwaite, D. 1997. Sustainable cities or cities that contribute to sustainable development? *Urban Studies* 34, no. 10: 1667–1691.

Saucier, M. 2008. Preparation underway to fix leak in Delaware Aqueduct. Department of Environmental Protection, City of New York Web site. http://www.nyc.gov/html/dep/html/press_releases/08-04pr.shtml.

Shrivastava, P. 1995. Environmental technologies and competitive advantage. *Strategic Management Journal* 16: 183–200.

Singh, B., and G. Sekhon. 1979. Nitrate pollution of groundwater from farm use of nitrogen fertilizers: A review. *Agriculture and Environment* 4: 207–225.

Smets, S., M. Kuper, J. Van Dam, and J. Feddes. 1997. Salinization and crop transpiration of irrigated fields in Pakistan's Punjab. *Agricultural Water Management* 35: 43–60.

Sonntag-O'Brien, V., and E. Usher. 2004. Financing options for renewable energy. *Renewable Energy* (May).

Srivastava, A., et al. 2009. The Great Lakes Collaboration Implementation Act of 2009: Final report policy analysis of HR 500. Columbia University, School of International and Public Affairs. http://www.columbia.edu/cu/mpaenvironment/pages/projects/summer2009/GreatLakeFinRept.pdf.

Starr, R. 1988. The region; A new campaign; Now, to get New Yorkers to turn off the faucet . . . " *New York Times* (May 8).

Sterman, J. 2009. A sober optimist's guide to sustainability [interview with Michael Hopkins]. *MIT Sloan Management Review.* http://sloanreview.mit.edu/beyond-green/a-sober-optimists-guide-to-sustainability.

Thomann, R. V. 1974. *Systems analysis and water quality management.* New York: McGraw-Hill.

Thrush, Glenn, and Carol E. Lee. 2010. Oil spill tests Obama's management style. *Politico.* http://www.politico.com/news/stories/0610/38161.html.

Timiraos, N. 2010. Fannie and Freddie resist loans for energy efficiency. *Wall Street Journal.* http://online.wsj.com/article/SB10001424052748704534904575132123115802584.html

UCS = Union of Concerned Scientists. 2007. Industrial agriculture: Features and policy. http://www.ucsusa.org/food_and_agriculture/science_and_impacts/impacts_industrial_agriculture/industrial-agriculture-features.html.

UNDP = United Nations Development Programme. 2005. Investing in development: A practical plan for achieving the millennium development goals. Office of the UN Secretary General. http://www.unmillenniumproject.org/documents/MainReportComplete-lowres.pdf.

UNEP = United Nations Environment Programme. 2002. Vital water graphics: An overview of the state of the world's fresh and marine waters. In *United Nations environment programme.* Nairobi: United Nations.

———. 2008. Organic agriculture and food security in Africa. Capacity-Building Task Force on Trade, Environment and Development. New York: United Nations. www.unctad.org/en/docs/ditcted200715_en.pdf.

———. 2009. Global trends in sustainable energy investment 2009. http://sefi.unep.org/fileadmin/media/sefi/docs/publications/Executive_Summary_2009_EN.pdf.

UNESCO = United Nations Educational, Scientific, and Cultural Organization. 2003. Water for people, water for life. In *World water assessment programme, UWWD report.* New York: UNESCO.

United Nations Food and Agriculture Organization. 1999. FAO factoids. http://www.fao.org/economic/ess/chartroom-and-factoids/chartroom/en.

Universidad Politecnica de Cataluña. 2007. Golf course irrigation with reclaimed water. Water Reuse Case Studies in Spain. http://www.ccbgi.org/docs/projectes_reutilitzacio/golf_course_irrigation_with_reclaimed_water_final.pdf.

U.S. Census Bureau. 2008. World population 1950–2050. http://www.census.gov/ipc/www/idb/worldpopgraph.php.

———. 2009. Metropolitan and micropolitan statistical areas. https://www.census.gov/population/www/metroareas/metroarea.html.

———. 2010. *International data base.* http://www.census.gov/ipc/www/idb/index.php.

USDA = U.S. Department of Agriculture. 2007. Census of agriculture shows growing diversity in U.S. farming. http://www.usda.gov/wps/portal/!ut/p/_s.7_0_A/7_0_1OB?contentidonly=true&contentid=2009/02/0036.xml.

———. 2009a. Farm structure. http://www.ers.usda.gov/Briefing/FarmStructure.

———. 2009b. Results and impacts for 1890 land-grant institutions programs. National Institute of Food and Agriculture Web site. http://www.nifa.usda.gov/nea/education/sri/multicultural_sri_1890.html.

U.S. DOE = U.S. Department of Energy. 2009. Home energy audits. Energy Savers Web site. http://www.energysavers.gov/your_home/energy_audits/index.cfm/mytopic=11160.

U.S. EPA = U.S. Environmental Protection Agency. 1997. The benefits and costs of the Clean Air Act: 1970 to 1990. Washington, D.C.: USEPA Office of Air and Radiation. http://www.epa.gov/air/sect812/copy.html.

———. 2003. Smart growth and heat islands. http://www.epa.gov/heatisland/resources/pdf/smartgrowthheatislands.pdf.

———. 2007. *Water Quality Handbook*. 2nd ed. U.S. EPA Web site. http://www.epa.gov/waterscience/standards/handbook.

———. 2008. Municipal solid waste generation, recycling, and disposal in the United States: Facts and figures for 2008. http://www.epa.gov/osw/nonhaz/municipal/pubs/msw2008rpt.pdf.

———. 2009. Heat island impacts. http://www.epa.gov/heatisland/impacts/index.htm.

———. 2009a. Water recycling and reuse: The environmental benefits. San Francisco: USEPA Water Division. http://www.epa.gov/region09/water/recycling.

———. 2009b. Environmental management systems. http://www.epa.gov/compliance/federalfacilities/compliance/ems.html.

———. 2010a. Case studies and key benefits of purchasing ENERGY STAR qualified products. Energy Star Web site. http://www.EnergyStar.gov/index.cfm?c=bulk_purchasing.bus_purchasing_key_benefits.

———. 2010b. About ENERGY STAR. Energy Star Web site. http://www.EnergyStar.gov/index.cfm?c=about.ab _index.

———. 2010c. How to conserve water and use it effectively. Nonpoint Source Pollution Web site. http://www.epa.gov/nps/chap3.html.

———. 2010d. Oceans, coasts, and estuaries: Invasive species. Office of Wetlands, Oceans, and Watersheds Web site. http://www.epa.gov/owow/invasive_species.

———. 2010e. Wastes—Resources conservation—Reduce, reuse, recycle. Office of Solid Waste Web site. http://www.epa.gov/osw/conserve/rrr/reduce.htm.

Walsh, D., S. Chillrud, J. Simpson, and R. Bopp. 2001. Refuse incinerator particulate emissions and combustion residues for New York City during the 20th century. *Environmental Science and Technology* 35, no. 12:2441–2447.

Water Board, City of New York. 2010. Rates and regulations. http://www.nyc.gov/html/nycwaterboard/html/rate_schedule/index.shtml.

Water-technology.net. 2010. New York City tunnel no. 3 construction, USA. Net Resources International. http://www.water-technology.net/projects/new_york.

Whitelaw, K. 2004. *ISO 14000 environmental systems handbook*. Amsterdam: Elsevier/Butterworth Heinemann, 2004.

Woody, T. 2010. Loan giants threaten energy efficiency programs. *New York Times*. http://www.nytimes.com/2010/07/01/business/energyenvironment/01solar.html?_r=2.

Woody, T., and C. Krauss. 2010. San Francisco area prepares for the electric car. *New York Times*. http://www.nytimes.com/2010/02/15/business/15electric.html?scp=2&sq=electric%20cars&st=cse.

World Bank. 2010. Water supply and sanitation. http://web.worldbank.org/WBSITE/EXTERNAL/TOPICS/EXTWAT/0,,contentMDK:21706928~menuPK:4602430~pagePK:148956~piPK:216618~theSitePK:4602123,00.html.

World Commission on Environment and Development. 1987. *Our common future*. Oxford: Oxford University Press.

World Health Organization. 2004. *Guidelines for drinking-water quality*. 3rd ed. Geneva: World Health Organization.

———. 2009. Progress on health-related millennium development goals (MDGs). http://www.who.int/mediacentre/factsheets/fs290/en/index.html.

WRI = World Resources Institute, United Nations Environment Programme, United Nations Development Programme, and the World Bank. 1996. World Resources 1996–97: The urban environment. http://archive.wri.org/page.cfm?id=929&z=?.

Zachary, G. P. 2008. Silicon Valley starts to turn its face to the sun. *New York Times* (February 17).

Zimmerman, J., J. Mihelcic, and J. Smith. 2008. Global stressors on water quality and quantity. *Environmental Science & Technology* 42, no. 12:4247–4254.

Index